Abstract

Fire services around the world share one common objective, to protect life and property. Although they have a variation of functions covered under their mission statements, the protection of life and property is a universal objective irrespective of nationality. However, the way this crucial objective is addressed from country to country, and even between fire services within the same country, varies significantly.

As such, there is little homogeneity in how fire services operate. Despite empirical evidence and a wealth of experience documented in books, manuals and training guides, establishing a broadly defined methodology towards effective emergency management and response, the status quo globally precludes the creation of an international benchmark for fire and rescue services.

This book aims to establish key performance factors against a set of baseline principles such as the standard of training and adoption of relevant science and technology, the existence of and conformity with standard operating procedures, operational health and safety standards, incident command structures etc. These elements, amongst others, will ultimately determine the overall performance of fire services.

The focus of this study has been on European fire services due to their inter-applicability of policies and processes relating to fire service operations. Comparisons with North American fire services for example become more complicated due to significant variations in their operational policies and practices. A comparative study between fire services in southern and northern Europe has been undertaken to establish key differences that would affect the quality of the service provided to their respective communities.

Acknowledgements

I would like to thank the below for their support and insightful contributions to this book:

David Alexander

Nikolaos Vythoulkas

Steve Apter

Ben Walker

Siemco Baaij

My Family

George Bogkias

Tonio Brix

Meisam Chengeni

Steve Dudeney

Bill Gough

Andrianos Gourbatsis

Paul Grimwood

Karel Lambert

Juan Carlos Campana Lopez

Pieter Maes

John McDonough

Hans Neiling

Shan Raffel

Michael Reick

Edward Savat

Stefan Svensson

Nikolaos Tsiplakidis

Mirella Ponte

Contents

Chapter 1 - Introduction

This book is derived from the author's PhD thesis and aims to understand how a universal benchmark standard of best practice could be developed and implemented by fire services to ensure that they are providing the highest level of service to the communities they serve. It further aims to explore and identify factors that strongly influence the organisational development and performance of fire and rescue services. The significant differences in how fire services deliver their core business have set the scope for this research. This is particularly the case between north European and south European fire services which are the focal point of this study. As such, the book aims to understand the social context, or rather, the socio-economic influences that would hinder a fire service from developing, restricting it from effectively achieving its fundamental goal of protecting life and property.

The study commences with a broad review of fire service practice on a global scale and negotiates matters such as the consolidation of benchmark standards, the proliferation of science and technology towards qualifying effectiveness and efficiency and how these have challenged more traditional modes of fire service practice, meaning that the way things that have traditionally been done by the fire service for decades might not necessarily be the best or correct way.

This study further draws upon the fire industry-wide problems experienced contemporality as a result of austerity measures that are requiring fire services to do more with less, restricting their ability to keep up to date with modern practices and potentially even sacrificing best practice standards to meet budget restrictions. It identifies that although there is consensus around the fundamental objectives required for protecting life and property, fire services achieve these goals in a variety of ways, meaning that there is a lack of homogeneity in fire service practice across the board, thus making it increasingly difficult to define a universal standard of best practice. Since achieving a platinum standard is so heavily reliant on funding, achieving it in many countries in the age of austerity, especially those in southern Europe that have been more heavily impacted, might not be reasonably achievable.

The selection of fire services from Portugal, Italy, Spain and Greece is central to This book as these are the countries that were most heavily affected by the financial crisis of 2008. These are countries that have similar characteristics in terms of geography and landscape and have cultural similarities in terms of family orientation and affiliation with religion etc. They are countries that rely heavily on tourism especially during the summer season and are generally known for their relaxed attitude to life when compared to north European countries. There are also similarities in terms of corruption and the influences this has on public sector organisations, politics, and other social and public policy matters.

This book also extensively analyses UK Fire Services as a control model against the sample study groups. There are many benefits to benchmarking against the UK model, most importantly, the availability of information. There also exists a national framework and a series of guidance documents that aim at promoting 'best practice' standards across all fire and rescue services in the UK. This provides a valuable starting point for understanding and comparing the operational structure of fire services.

When embarking on this study, the extent to which the fire services operated in Italy, Spain and Portugal was unknown. There was a general impression acquired from discussions with firefighters from these regions and from reviewing online videos that, although problematic when compared to their north

European counterparts, they were in fact making slow but steady progress even through times where the financial crisis was at its peak. Conversely, the author was familiar with how the fire service worked in Greece. For a European country, there were significant concerns about how the Hellenic Fire Service conducted its day-to-day business. There are numerous positive and noteworthy examples of the Hellenic Fire Service that is represented by many experienced and competent firefighters. However, the general culture of the fire service (as an organisation) in Greece is unequivocally poor. An in-depth analysis of a case study of a firefighter fatality in Greece in 2014 serves as the focal point of this study as it details the systemic failures that led to this tragic outcome and that still persist today.

The analysis of this case study was fundamental in developing the principal questions raised in This book. These questions aimed at identifying a broad consensus around ensuring quality assurance for service delivery of fire services on a global scale and looked at how such organisations in southern Europe performed in relation to a broad set of values when considering the limitations and restrictions presented by the financial crisis and other social or cultural factors. It also aimed at identifying the differences between Greece and the remaining three south European countries in this study, seeing as there was already a pre-existing understanding and in-depth knowledge around the culture of the Hellenic Fire Service but little understanding of the intrinsic matters and culture pertinent to the fire services in the other sample countries. Finally, the thesis aims to understand, where differences restricting fire service development were identified, whether there were any common causal factors that could be attributed to this situation and, most importantly, how these could be addressed.

In his 25 years of service as a volunteer firefighter, the author has had the opportunity to train with fire services from around the world and has studied them extensively. Through this, however, it slowly became apparent that Greece remains far behind most north European fire services, especially in terms of defined processes and systems that serve to ensure their efficiency and effectiveness as emergency services. This intrigued the author to understand why this might be so. The simplest explanation was of course the financial crisis that featured so prominently at the time of writing. It is understandable that without sufficient funds, progress will be hindered, old equipment cannot be replaced for new and, as a result, the quality of service would likely suffer. However, Greece was not the only country in financial crisis at this time. In the same broad geographical region, Italy, Spain and Portugal were also suffering from the impacts of austerity. These countries were also similar to Greece in that they presented significant issues around clientelist relationships, the influence of politics at all levels of organisational structures and society alike, nepotism, favouritism and corruption. Nonetheless, the fire services in these countries still appeared to be progressing, perhaps not to the extent of other countries like Belgium, the Netherlands or the UK etc., but nonetheless they appeared to be making advancements as emergency response organisations.

An initial hypothesis was that other countries in southern Europe with similar characteristics to Greece were progressing despite the financial crisis. These findings will be assessed through the data analysis section of This book. A second hypothesis of this research was that most north European fire services were working to national standards that provided assurances in terms of service delivery, operational homogeneity and interoperability. This hypothesis proved to be incorrect.

However, north European fire services experience intrinsic issues that are contextual to their organisation and culture and these problems irrefutably impact on the quality of service they provide to their respective communities. In the UK, for example, there is a drive to establish smaller crew sizes with small rapid response vehicles to attend lower risk incidents, rather than send out large fire engines with five-person crews to every incident. Though this could work in principle, other fire services accept smaller staffing levels on frontline fire engines as standard practice despite research demonstrating the greater risks presented to both fire crews and the public alike (Grierson and Kirk, 2022). In the Netherlands, the fire service is experiencing a steady decline in the recruitment of part-time firefighters which is problematic for the entire civil protection mechanism considering that over 80% of the fire service is staffed by part-time firefighters (Nederlands Instituut Publieke Veiligheid, 2019: p -7). In Germany, there is no national standard for training meaning that every regional fire service can regulate its own training and operational procedures (Platt, 2018).

It is understandable that fire services will operate differently in some cases due to the intrinsic characteristics and risks present in their local areas as well as their capacity to deliver fire and rescue services, their funding and demographic factors – amongst others. However, the lack of consensus despite scientific research and other experience-based evidence means that the work of fire services is undertaken very differently across the world. A baseline of common principles which most fire services should adhere to has been determined. These have to do with the protection of firefighters through the provision of suitable personal protective equipment, the creation and use of standard operating procedures and guidelines that inform operational tactics, a commitment to training and development, a defined incident command structure and the delegation of roles and responsibilities through rank etc. Unlike their south European counterparts that appear to be addressing these matters with a gradual consolidation of such principles into their organisational culture and which have arguably assisted them in their progress, the Hellenic Fire Service is yet to achieve an appropriate level of commitment to many of these basic principles, nor has it fostered a culture around professional development.

Through this research, there is an attempt to determine the impact socio-economic constraints and other cultural influences have on fire service development. It is also attempted to understand what these factors are, the extent to which they affect fire service development, and any interesting patterns that can be identified between the sample groups. In addition, it is attempted to determine whether the Hellenic Fire Service is indeed an independent case with notable differences from the other sample groups of south European fire services. This research also attempts to establish differences between north and south European fire services where it has been hypothesised that north European fire services will demonstrate greater organisational structures and assimilation to baseline principles than south European fire services. The purpose of this study is to establish the possibility of further research initiatives and possible strategies to address performance and organisational issues for fire services in an attempt to consolidate consensus against a set of baseline standards and provide assurances in how fire services protect their communities.

The literature review utilises a broad range of formal and informal research and writings to set the context of this study. This section aims to give context to the problems This book attempts to address. The first part of the literature review (Chapter 2) links boarder research with the context of the fire service organisation such as organisational learning theory, organisational culture etc. As the Hellenic Fire Service sits at the centre of this study due to its unique profile and negative bias in the data set, the last section explores the socio-economic and political history of Greece in order to understand key factors that could explain this uniqueness. Chapter 3 then explores the more intrinsically operational elements of fire service practice and identifies key issues affecting safe and effective practices such as heavy reliance on experience rather than science and other failures within the operational structure such as crew inconsistency, a lack of, or poor, training, fluid incident command structure, an absence of understanding on the importance of radio communications, personal protective equipment and so on etc. Chapter 4 is a case study of a fire fighter fatality that occurred in Athens, Greece in 2014. This case study clearly demonstrates and affirms the likely outcomes of inconsistent and weak organisational and operational structures within the fire service. It further, allows for a clearer understanding of the status quo of the Hellenic Fire Service to provide greater context when analysing the results of the dataset. Chapter 5 is the methodology employed in this study in both qualitative and quantitative format. It determines the research problems that are to be addressed and the output data arising from the questionnaire. It further discusses the validity of the qualitative research used more broadly through the body of the thesis. Chapter 6 analyse the results of the data set and Chapter 7 is the discussion relating to those data and how they are relevant in answering the research questions.

This book set out to explore the feasibility of developing and implementing a universal benchmark standard of best practice for fire and rescue services, with a particular focus on the contrasting practices and conditions between northern and southern Europe. By examining the organisational development and performance of fire services in Portugal, Italy, Spain, and Greece, and using the UK Fire and Rescue Services as a comparative control model, the research has identified a range of critical factors that influence service delivery—both positively and negatively.

The findings reveal that while all fire services fundamentally share the objective of protecting life and property, their ability to achieve this goal is influenced by a complex interplay of economic, political, social, and cultural factors. South European fire services, though affected by the austerity measures following the 2008 financial crisis, have shown incremental progress in modernising their operational practices – at least this can be surmised from the data. The Hellenic Fire Service, however, has demonstrated more entrenched systemic issues, where cultural resistance, organisational inefficiencies, and lack of professional development continue to hinder performance.

Furthermore, the research disproved the hypothesis that north European fire services operate under uniformly standardised practices. Despite having access to more robust funding models and frameworks, these services also face internal challenges such as staffing shortages, inconsistent training, and operational decentralisation. This suggests that the absence of a unified European standard for fire service operations is not just a south European issue but a broader systemic challenge.

Through both qualitative and quantitative data, including a case study of a firefighter fatality in Greece, This book has demonstrated the real-world implications of organisational shortcomings. It has also emphasised the importance of standardised training, incident command systems, personal protective equipment, and a culture of continuous improvement. The central conclusion of this study is that while a universal benchmark standard is desirable—and in many ways necessary—its implementation must account for national and regional contexts. A flexible, principles-based framework that allows for local adaptation, but which is rooted in evidence-based best practices, may offer the most effective path forward. The study also underlines the need for continued research and cross-border collaboration to ensure that fire and rescue services can respond effectively to both traditional and emerging risks in that are increasingly interconnected in modern society.

By identifying the gaps, barriers, and enablers of fire service performance, this research provides a foundation for further inquiry into strategies for improvement and standardisation. Ultimately, it reinforces the idea that excellence in emergency services is not only a matter of resources but of leadership, accountability, and a commitment to professional integrity.

Chapter 2 – Literature Review

An issue of utmost importance that must be highlighted here, is that there is not extensive available literature about fire services from a social sciences perspective, and the bulk of existing works are documented by fire service practitioners with extensive experience who have acquired general acceptance from the industry. Unfortunately, though most of these writings can be seen as accurate, insightful and authoritative, in reality they often lack the theoretical and empirical foundations to achieve academic or scientific acclaim. Conversely, there is quite extensive literature and research on technical aspects relating to the fire service, their structure, and operations. This type of literature has been essential in enabling the establishment of a performance benchmark on which to assess fire services. The literature review can be seen as drawing information from three main sources.

The first source would be in respect of a review of the technical literature that is currently available. This is fundamental to understand why fire services do what they do and the variety of methods which they employ to achieve their organisational goals. The second source of information would be relevant to the more academically limited literature arising from a social sciences perspective on matters such as fire service culture and the intrinsic characteristics of these organisations that make them unique but generally quite resistant to change. The third source would be drawn from a more general overview of relevant literature on matters such as culture and organisational change, risk perception, organisational resilience, and organisational learning theory, political history, corruption etc. This aspect of the literature review will almost serve as the structural supports to this study to provide a broader understanding of the underlying themes that can be directly transferred, applied and reviewed in a fire service context.

It has been found that there are some broadly agreed upon denominators that are shared amongst most fire services on a global scale that underpin their professionalism as emergency services. Although, as mentioned earlier, fire services tend to achieve the same fundamental goals for protecting life and property by using a variety of different methods, there appears to be a general consensus amongst most fire services in respect of those denominators. These refer to matters such as, the importance of minimum staffing levels and first response configurations, training, the establishment of standard operating procedures or guidelines, a structured incident command system, a stringent health and safety policy, radio communications, accountability processes and so on. This can be referred to as a 'systems-based' approach to organising and managing fire service operations. By analysing relevant technical literature and case studies around these key areas it became possible to establish a set of basic principles under which to assess the fire services. Indeed, the structure of the questionnaire used in the quantitative part of this study was designed such, so as to create a profile for each studied fire service against those basic principles. This would essentially assist in developing a metric for measuring their efficiency and effectiveness in achieving their fundamental goals for protecting life and property.

Fire services globally are considered as primary response organisations that attend the vast majority of emergency events (Hart, 2011). As society has become more technologically advanced, the traditional role of the fire service simply 'fighting fires' has changed significantly. Fire services around the world do indeed put out fires, however, they also attend road traffic collisions, flood incidents; free people trapped in lifts; perform animal rescues (Murphy, 2014: p 37); the list is extensive and will vary from country to country. For example, some fire services (e.g. Hamburg, Dublin, Brussels, numerous states in

North America, Canada etc.) have the ambulance service embedded into their organisational structures. Others have specialised Urban Search and Rescue Teams to deal with earthquakes and building collapses etc., Hazardous Materials Response Teams that are called to attend Chemical, Biological, Radiological, Nuclear and high yield explosive (CBRNe) incidents, specialist multi-agency anti-terrorist units etc[1]. There are also many that place great emphasis on community safety and fire prevention initiatives that form part of their daily function, with visits and training provided to the more vulnerable members of the community as well as local businesses. The modern firefighter's role can be extremely diverse.

This diversity is not only apparent in the ever increasing role of the firefighter, but also in the quality in which these services are provided to the public, which can vary significantly. There are several points that must be addressed here. The first has to do with how performance is measured. If a fire in a building is extinguished, this is usually perceived as a success as it is based on outcome alone. Similarly, if a fire in a house is extinguished but the occupant is not saved, this might be perceived as a failure. In any emergency intervention, outcome alone cannot be viewed as the only source for measuring performance due to the dynamic nature of these events and the likelihood of what 'could' have gone wrong (or better) if circumstances were different. Therefore, it must be determined how fire services can measure their performance with the intention of ensuring that they maintain the highest standard of service to their communities whilst simultaneously protecting the health and safety of their firefighters. It has been determined through this research, that there is great variability in how fire services deliver their core business. In order to understand this, it is

---

[1] https://www.london.gov.uk/who-we-are/what-london-assembly-does/questions-mayor/find-an-answer/lfb-training-support-multi-agency-response-terrorist-attack

important to identify how organisations learn and how this is applied to the fire and rescue service.

Organisational Learning

One critical area to explore, is 'organisational learning theory', which can assist in establishing the critical processes that help organisations such as the fire service learn and develop. Though there are numerous definitions of 'organisational learning' it is argued that it is a change in the organisation's knowledge that occurs as a function of experience (Argot & Spektor, 2011: p 4) and that experience can be acquired on novel tasks or on tasks that have been performed repeatedly in the past (Katila & Ahuja, 2002; March, 1991). However, though this might be true, there is a much broader spectrum of variables that will influence the extent of learning.

The first is whether the organisation is culturally open to learning and, secondly, to what extent it is capable of understanding the necessity for change. This is especially important in cases where certain actions are used to address a given problem, in most cases successfully, but where these actions are far superseded by better and more efficient methods yet not adopted or practiced. It is commonly understood that firefighters learn by application, repetition and experience (Sandlebeen, 2002; Denef, Ramirez, Dyrks and Stevens, 2008) and if what they do is working, then they will often continue to use that method as standard practice. Chief of Whidbey Island Fire Department, Ed Hartin (2013) further attests that '*[fire service] strategies, tactics, and doctrine must be evidence based, must have a sound theoretical foundation and be supported by both scientific research and practical experience*'. However, he goes on, that, '*unfortunately, this profession continues to struggle to integrate these elements and is saddled with conclusions based on experience without understanding*'. It appears that this limitation is not specific to fire services alone and is understood as a de-facto characteristic across the field of organisational learning theory altogether. Choo (1996: p-330) argues that organisations find it difficult to unlearn their past and question inherited assumptions and beliefs, to reject existing practices as the only viable alternatives. Indeed, Senge (1990) has also warned that many organisations are unable to fully function as knowledge-based organisations as they suffer from 'learning disabilities.' To overcome these disabilities, it is purported that the learning organisation must develop the capacity for both generative and adaptive learning (Ibid). Learning, or knowledge, is also achieved through an environmental context, something which refers to external factors occurring in the cosmos that would have the potential to influence an organisation's main functions in one way or another (Fiol & Lyles, 1985).

Although Argot and Spektor (2011) argue that individual learning is necessary, they imply that it is not sufficient for group or organisational learning. Elkjaer (2003: p-2), believes that it is individuals who acquire specific skills, information, and knowledge in order for organisations to learn and to become 'Learning Organisations'. Organisations comprise people and people will serve as the primary receiver (or suppository) of knowledge and information. It is then what happens with this knowledge that will determine a learning outcome.

Therefore, the further utilisation of knowledge into a driver that can facilitate learning at an organisational level is fundamental. This could be seen as a critical interaction between information and the capacity and adequacy of available systems and processes of an organisation to interpret, structure and disseminate it effectively. This means that knowledge can often become futile if there is no method of capturing it, giving it structure, meaning and dimension (beyond its original source). This point is further affirmed by Fiol & Lyles (1985: p-804) who suggest that there are four contextual factors that affect the probability that learning will occur; a corporate (or organisational) culture conducive to learning, a strategy that allows flexibility, an organisational structure that allows both innovativeness and new insights, and the environment.

Fire services in Northern Europe and developed Western countries generally appear to understand this concept and promote this position through their organisational culture. In the 'Good Practice Guide for Fire and Rescue Services' of the National Operational Guidance Program (2016: p 4) in the United Kingdom, the importance of learning is expressed as follows:

> *"Knowledge learned from operational activity should be shared with all fire and rescue services, and the wider sector where appropriate, to help reduce risk. A consistent and shareable learning process will enable fire and rescue services to improve performance at incidents, cope with future changes in the operational environment, develop and innovate to provide the best service to the public and improve the safety of firefighters".*

Chief Scott Thompson of the Colony Fire Department in Texas (2014), in his article 'Training Basics and Essentials for the Fire Service' also argued that *'when we [the fire department] fight a fire from which nothing is learned, that fire is merely an event. Training that does not connect to reality is useless'.* Inadvertently, this statement reflects the need to establish deliberate processes around learning and that in the absence of these, events become meaningless.

It is now understood that knowledge, in the first instance, is acquired by individuals representing an organisation. Choo (1996: p-331) argues that that humans are not ideal drivers for processing and sharing information in a way that will promote organisational learning. Choo refers to organisations as decision-making systems stating that *"in the decision-making view, the essential features of organisational structure and function may be derived from the characteristics of human decision-making processes and rational human choice".* However, he continues, that this concept is idealistic, as rational choice would require a complete search of available alternatives, reliable information about their consequences, and consistent preferences to evaluate their outcomes, and in the real world, such demands on information gathering and processing are unrealistic (Ibid). In attempting to qualify the limitations of effective human decision-making, Herbert Simon (1957: p 198) suggested that decision-making in organisations is constrained by the principle of 'bounded rationality' which is described as follows:

> *"The capacity of the human mind for formulating and solving complex problems is very small compared with the size of the problems whose solution is required for objectively rational behaviour in the real world, or even for a reasonable approximation to such objective rationality."*

Simon (1976) later identified a series of boundaries that limit the capacity of the human mind for rational decision making. These are separated into three categories. The individual is limited:

1. By his mental skills, habits, and reflexes;
2. By the extent of knowledge and information possessed, and;
3. By values or conceptions of purpose which may diverge from organisational goals.

As a result of these limitations, it is organisations that become central in shaping or controlling decision-making processes through the creation of overarching goals, providing the tools required for achieving them and investing in its members and their training to enable them to bring those goals to fruition. Simon further asserts that the organisation influences its members' behaviours by controlling the decision premises upon which decisions are made, rather than controlling the actual decisions themselves. In organisational learning theory, organisations are seen as rational systems with a greater level of reliability than its individual members. Organisations overcome this by designing decision premises and decision routines that guide or control individual decision behaviour. These are based on specific tasks in a process-driven format that are designed to achieve the overarching goals and objectives of the organisation. Goals and objectives are set first, and when participants encounter problems in the pursuit of these objectives, a search for information on alternatives and consequences takes place, followed by evaluation of the outcomes according to the objectives and preferences (Choo, 1996: p-332).

Organisational Learning in the Fire Service

The fire service is a constantly evolving profession that must keep pace with advancements in technology across multiple functions in order to ensure the provision of an effective service to the public. Most activities relating to the fire service have changed quite significantly, even from just ten or 20 years ago. This is, of course, as a result of greater capabilities provided through newer and better equipment designed specifically to overcome difficulties experienced in the past. This is also due to the implementation of better fire safety legislation and building design standards, the availability of life safety systems such as automatic fire detection, sprinklers and the promotion of public awareness campaigns around fire safety etc. since the 1980s that have shown a consistent decline in the number of emergency incidents attended by fire and rescue services in many countries (U.S. Fire Administration, 2020: p – 2; Hewitt et al., 2002; Roguski & Szułczyńska, 2021). Conversely, the introduction of new technologies such as Lithium Ion batteries in electric vehicles (Bisschop R, Willstrand O, Amon F & Rosendgren M, 2019: p - 45) the proliferation of synthetic materials and plastics in domestic products (Kerber, 2011) as well as the impact of climate change (CTIF International Association of Fire and Rescue Services, 2023) have introduced significant challenges to which fire and rescue services have had to adapt.

Learning through experience and adapting to change is a fundamental basis for fire service development. However, there appears to be a prominent antithesis between fire services in countries that experience greater political, financial and social stability. In general, it appears that fire services in these countries with these requisites have a different capacity around learning, implying that there must be a difference in 'context' (Glynn, Lant & Milliken, 1994). Kane et al., (2005) argue that a context where members share a superordinate identity has been found to lead to greater knowledge transfer. Similarly, contexts where members trust each other (Levin & Cross, 2004) or feel psychologically safe (Edmondson, 1999) have been found to promote organisational learning. In addition to this, a "learning" orientation has been shown to facilitate group learning up to a point (Bunderson & Sutcliff, 2003) whilst a culture of psychological safety (Edmondson, 1999) that lacks defensive routines (Argyris & Schon, 1978) has been found to better facilitate learning.

It has been found that specialist organisations learn more from experience than generalist organisations (Ingram & Baum, 1997; Haunschild & Sullivan, 2002). The fire service would fall under the category of a specialist organisation with the added complication of frequently performing its function under difficult and demanding conditions with a significant element of risk. Although it is established that there is a notable level of diversity between fire services from around the world, many in the more developed countries appear to share commonalties in the ethos around achieving service excellence, though the methodology on how these are achieved between each service can vary and, in some cases, significantly. The main difference between fire services in northern Europe and southern Europe is the absence of baseline standards for the latter. These services appear to rely more heavily on experience, and what has worked in the past is often recycled and assumed as standard practice. In such cases, organisations will find it more difficult to instil a learning culture and this will fundamentally have to do with 'capacity' across a series of different dimensions (funding, ability, adaptivity, understanding etc.) (Sukuco et al., 2021).

A U.S. Fire Administration Technical Report 89 (1998: p - 4), 'Special Report: The Aftermath of Firefighter Fatality Incidents: Preparing for the worst' argued that firefighter fatalities often lead to significant changes in operational procedures. This is a commonly shared understanding and highlighted in numerous documents, for example, the UK Fire Brigade Union's (2015) report 'Firefighter Fatalities at fires in the UK, 2004 – 2013: Voices from the Fireground' and the National Operational Learning document 'Good Practice Guide for Fire and Rescue Services' (2016), which is a guidance document to ensure that fire and rescue services are identifying cues to effect change as and when required in light of service effectiveness.

Adam Course of Avon Fire and Rescue also developed an international database through the Institution of Fire Engineers portal of line of duty injuries and deaths case studies. The UK fire services in particular (but also the USA) have provided investigation reports into most near-miss or firefighter fatality incidents that have occurred in the UK since the King's Cross fire. These reports (Blaina 1996, Bethnal Green 2004, Shirley Towers

2010, Grenfell Tower 2017, amongst others) provide important insight into errors that have occurred allowing for affirmative action to avoid the recurrence of similar errors. Similarly, Karel Lambert of the Brussels Fire Department has developed a 'Learning from Incidents' section on the website www.cfbt-be.com that documents case studies of firefighter injuries and fatalities from around the world. Unfortunately, it is known that this does not apply universally, even in more developed countries and, as a result, important learning opportunities are missed. The consequences of such missed opportunities are highlighted in the Mantzios case study discussed later in This book.

Change in the Fire and Rescue Service

With learning comes change and fire services organisations (like many other organisations) have historically been resistant to change (Napp, 2015: p 5) because they feel that change is an indictment of prior decision or actions (Jick and Peiperl, 2003). Fires and other emergency incidents were dealt with differently in the past. Methods, tactics and equipment have been updated and modified in-line with industry developments and many fire services around the world, including the supporting industry, are at the top of their game pioneering in the field of operational fire service tactics and firefighter safety. However, it regretfully appears that many countries have remained committed to tradition without questioning their effectiveness or examining other alternatives for achieving service excellence and, consequently, firefighter safety. Although, as aforementioned, UK fire services have been consistent in investigating high profile incidents, this does not automatically result in the desired change. This would indicate weaknesses in either an ability to learn[2], or to understand the context of critical errors being investigated.

---

[2] Difficulties in learning and change can indicate dominant cultural influences (Kilman et al., 1986).

An inquest into the Lakanal House fire in London in 2009 where 6 people lost their lives revealed a series of critical failures. These had to do with issues around the installation of inappropriate cladding for tall buildings (i.e. over 18m in height), significant compartmentation issues that allowed smoke and fire to spread throughout the building and a failure in the performance of fire doors. In addition, it was found that the London Fire Brigade response was problematic due to a miscommunication of information between control room and on-scene fire crews, a lack of understanding of the building layout, operational training issues and, most importantly, a failure of control room staff to critically assess the effectiveness of the stay-put policy and the conditions under which this advice should have been abandoned (London Fire and Emergency Planning Authority, 2013). A failure which ultimately led to the deaths of the six people that failed to evacuate due to incorrect advice. The coroner's inquest put forward a series of recommendations to overcome issues identified as contributing to the tragic outcome of the Lakanal House fire. These incorporated recommendations both around the provision of better fire safety measures and controls, as well as procedural and training improvements for the London Fire Brigade.

Seven years on, the UK experienced it's most devastating fire since World War II, resulting in 72 fatalities at the Grenfell Tower in London on 14th June 2017. Once again, the critical failures identified in the ensuing four-year inquiry were the presence of highly combustible cladding affixed to the building and the absence of cavity barriers coupled with combustible insulation that allowed the fire to spread unfathomably up the face of the building and spreading into multiple flats. There were significant issues also identified in the London Fire Brigade's response. Sir Martin Moore Bick, Chair of the Inquiry stated in 2019 that *"the LFB was not prepared for a fire of this scale and complexity. Serious shortcomings in its response were exposed, particularly in relation to incident command, control room operations, and decision-making under pressure."* These issues can be summarised as follows:

- A lack of training in how to recognise the need for an evacuation or how to organise one.
- Incident commanders of relatively junior rank being unable to change strategy.
- Control room officers lacking training on when to advise callers to evacuate.
- An assumption that crews would reach callers, resulting in assurances which were not well founded.
- Communication between the control room and those on the ground being improvised, uncertain and prone to error.
- A lack of an organised way to share information within the control room, meaning officers had "no overall picture of the speed or pattern of fire spread.

(Manning, 2019)

It is apparent, that many of the key failures identified from Lakanal House were repeated in the Grenfell Tower Fire, despite the recommendations arising from the coroner's inquest. In addition, it was also noted that a key issue affecting the evacuation of residents from the upper-floors was as a result of lobby doors to the protected stair not being closed properly to allow for firefighting hoses to pass through (Grenfell Tower Inquiry, 2019). It is commonplace for fire services (especially in the UK) to establish a safe staging area from where to commence firefighting operations. This is called the bridgehead and is usually located one or two floors below the fire floor (London Fire Brigade, 2024). However, the result of this strategy is that in order to extend a hose onto the fire floor, doors protecting means of escape such as lobby doors will not fully close allowing toxic smoke to enter the stair enclosure. In such an event, stair conditions may become untenable and cannot be used for escape if the stay-put policy needs to be overridden – as was the case at Grenfell Tower (Galea, 2019).

For over a decade, fire services in Germany and mainland Europe have been successfully using the 'smoke stopper' to counter this problem (Ibid). Developed by Dr Michael Reick, a regional fire chief in Germany and former leader of the fire research laboratory at the University of Stuttgart, the smoke stopper is a simple device consisting of a fire curtain fitted onto an adjustable framing system that can be pushed up into the top of a door frame and fixed in place. The curtain is then released which drops to floor level allowing hoses to be pulled through the bottom part of the curtain whilst restricting smoke spread onto the means of escape (Lambert, 2015). It could be argued that the deployment of smoke stoppers as a tactical decision at Grenfell Tower to protect smoke ingress into the stair might have potentially reduced the number of fatalities, although this can only be speculated. Despite the truly heroic efforts of the London Fire Brigade firefighters at Grenfell Tower, the organisations failure to learn from previous errors and upgrade their operational procedures for stair protection systems, in line with those used in by fire services in mainland Europe (and some fire services in the UK), left them open to heavy criticism.

Jim Crawford (2013), a retired Deputy Chief of the Vancouver Fire Department, discusses a cultural context within the fire service relating to organisational change that supports the opinion that 'if it isn't broken, then why fix it?' Indeed, Grimwood (2008) emphasises this point in the introduction of his book 'Euro-firefighter' where he mentions that in the days before health and safety and managing risk on the fire-ground became commonplace, firefighters seemed to get things done just fine without injuries but also meeting mission objectives. Therefore, if fundamental mission objectives are being satisfied then perceptions relating to learning opportunities and change will likely be overlooked. Crawford's point, however, was that fire services, by the very nature of their work in society, will make them susceptible to change. He argues that *'fire services should look at making cultural changes to survive, to thrive and meet new challenges (like the ones that ask them to do more with less)'*. In other words, they must be able to adapt to changes to ensure resilience as an organisation. However, Grimwood (2017: p 1) argues that fire services adaptation to change to achieve fundamental service delivery goals rarely occurs during global economic downturns. The financial crisis (originating circa 2008) has had a global impact on fire services to more or lesser extent, and it is by virtue of this circumstance that fire service development will be and has been limited (National Audit Office, 2015).

In summary, it has been established that fire service development and organisational learning is critical if fire services are to be able to provide an adequate service to the communities they serve. Organisational learning is influenced by a number of factors to include culture, ability to learn, availability and access to funds, identifying opportunities to learn, develop and change and the promotion of an ethos around excellence. Through this research, it appears that there is a greater inclination towards poor fire service practice in countries that have greater social, political and economic instability. This was apparent in the comparisons between north European and south European fire services researched in this study.

However, examples of bad practice are not just restricted to these less advantaged countries. As the work of fire services from around the world, though broadly established to meet the same fundamental objectives of protecting life and property, achieve this goal (to a greater or lesser extent) in a myriad of ways, this would suggest that organisational learning is not unilateral and, in fact, highly contextual. The challenge is identifying where the benchmark lies that should be met by every fire service regardless of their structure and financial status. If this is established, then consideration must be given as to whether this is indeed achievable and, if not, what the underlying causes are, if and how they can be overcome.

The Impact of Data Analysis on Fire Service Development

Chinese military leader, strategist and philosopher in his writings 'The Art of War '(5th Century BC) purported that *"If you know your enemy and know yourself, you need not fear the result of a hundred battles. If you know yourself but not your enemy, for every victory gained you will also suffer a defeat. If you neither know your enemy or yourself, you will succumb in every battle"*. The structuring of fire services requires significant planning to ensure that communities are provided with adequate protection against hazards and their associated risks. A fire service cannot provide effective emergency cover without understanding the intrinsic qualitative and quantitative risk indicators that are pertinent to the areas they aim to protect. These indicators need to be assessed to determine their value so that the best possible structure, ensuring that the fire service can adequately meet their objectives for protecting life and property, is established. This is itself a fundamental basis for fire service development.

Since immemorial, mankind has sought ways to effectively reduce the impact of certain hazards that were found to present a risk to life and livelihood. As time has moved on, this task has moved out of the hands of the lay person and gradually been bestowed upon professional bodies, especially in most developed countries. Fire services have thus transformed from their original forms into professional public service organisations with often new and innovative ways of overcoming challenges faced in the line of duty. Fire services must now qualify why they do what they do, what resources they need, how they will allocate these, the level of training that must be provided to their staff and have a clear understanding of the extent and types of risks they are likely to be called to in an emergency. Fire services cannot make assumptions about such matters because if they do, it is unlikely that the level of service they will provide will be adequate in protecting the communities they serve. Without sufficient and reliable information, it is unlikely that the structure of a fire service will be in a position to effectively meet its objectives. It will provide an overtly reactive service that will undoubtedly be less effective than a proactive organisation with specific targets and mission objectives.

Fire services intrinsically produce (or at least should produce) a plethora of information. They also require data from numerous sources to assist in their daily operation. The extent of this information has the ability to be so extensive that it can also be referred to, within context, as Big Data. 'Big Data' refers to extremely large data sets that can be analysed to reveal patterns, trends and various other insights (Sun et al,. 2018). In a fire service context, data can be used to prevent fires, protect firefighters, and better allocate resources (Markley, 2016), amongst other things. Fire services were initially established to deal with emergencies related to fire. If the problem of fire is to be challenged effectively though, they would need to understand the causes of fire, where they were most likely to occur and who and what was most at risk. This information would be able to assist in reducing the likelihood of fire occurring through prevention and protection initiatives, the extent and type of resources required to effectively respond to possible fire incidents, and the geographical location of fire stations to minimise response times to have the ability to tackle fires in their early stages.

One of the principal philosophies now adopted by many fire services is that adequate prevention methods will significantly reduce the need for intervention. This was demonstrated through the analysis of incident data where an increase in prevention initiatives (funding, campaigns) was associated with a significant reduction in the number of emergency incidents attended by fire services (Hewitt et al. 2002). Such efforts are often strongly supported through government policies and regulations. Fire service interventions are costly and put the public as well as firefighters and other emergency workers at risk. The damages that occur as a result of an emergency can result in loss of life, injuries, a loss of livelihood, damage to property and infrastructures. They also have a knock-on effect that can lead to unemployment, homelessness, depression and other mental health issues, situational displacement etc. (Ritchie et al., 2022). In relation to fires, the National Fire Chiefs Council (2015: p 5) reported that in 2004, the estimated cost of fire deaths in the UK was £1.65 million for each death. Additionally, it was found that 'the total economic cost of fire in the UK was estimated at £7.03 billion, equivalent to almost one per cent of the gross value added of the economy. Of that cost, £2.5 billion can be attributed to the consequential costs of fire such as property damage, lost business and the loss to the economy from injuries and deaths. By 2008, this estimated cost had increased to £8.3 billion'.

A later report issued by the Local Government Association (2017: p 15) identified that it was the Fire and Rescue Services Act, consolidated in 2004, that introduced a statutory duty to promote fire safety that was ultimately responsible for the significant reduction in fires and fire related deaths from 2004 onwards. This law, combined with the removal of a perverse funding arrangement (where the more fires that were attended one year, the more funding Fire Rescue Authorities received from government the next), shifted thinking on how to better manage community risk. Rather than doing so primarily by firefighters attending and mitigating the impact of fires and other emergencies, the focus shifted towards working to prevent emergencies from happening in the first place. This move towards prevention in favour of emergency response was fundamental in reducing the number of emergencies attended in the UK by two-thirds and halving annual fire death figures in less than fifteen years. Figure 1 clearly demonstrates the rapid decline in the total number of fires in the UK since the implementation of the Fire and Rescue Services Act in 2004. Since then, there has been a consistent decline in the total number of fires year upon year and remaining consistently low, thus presenting a strong case for the promotion and financing of prevention initiatives over emergency response.

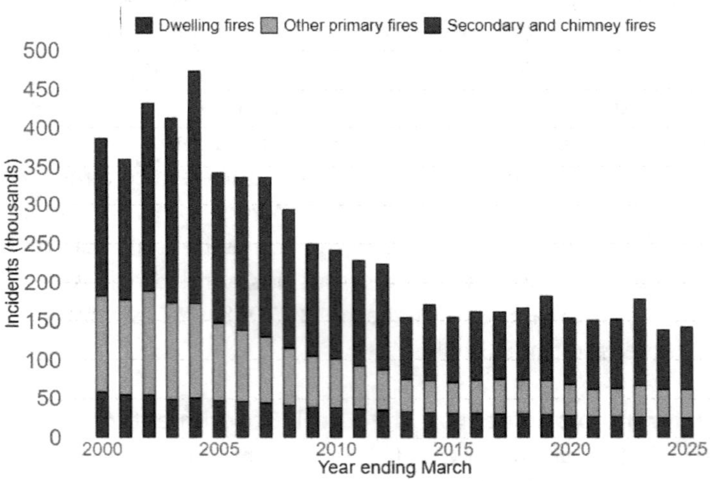

Figure 1: Total fires attended by type of fire, England 2000 to 2025

(Home Office, 2025)

As previously mentioned, the exchange and analysis of information has always been a fundamental part of fire service development, even in its more basic forms. In their article in online portal, 'Fire Rescue Magazine', Quinn and Roden (2014) write about how older firefighters are able, with startling accuracy, to predict when and where a fire will next occur. Though informally, this is often referred to as a 'hunch' or 'gut feeling', in reality, their guess is based on an assessment of a wealth of past experiences applied to present conditions. This is a very basic form of information analysis that nonetheless acquires a level of reliability due to the frequency and subsequent likelihood of that same event occurring at given points in time. The scope of data analysis is to convert this informal knowledge base into meaningful interpretations that will assist in the establishment of organisational priorities. As stated by Amanda Gardiner of business analytics specialists 'SAS' who have designed Risk Analysis software for fire services; *"..understanding risk allows you to target resources. What the analytics does is it takes that gut feeling away and it has true metrics applied to it that people can use within the business."*

The potential level of insight that can be acquired by a modern metropolitan fire service is phenomenal. This means that the ability to tailor service delivery to satisfy specific organisational objectives is immense and supported by firm evidential bases. It is now suggested that by risk profiling developed through Big Data analysis, fire services will even be able to predict future emergency events, such as fires, and be in the position to address these proactively (Graham, 2013). This is a significant cornerstone in the history of fire service development that is likely to influence a drastic change in fire service culture globally moving away from traditional reactive models of service delivery towards a more proactive approach focused on prevention. Many large metropolitan fire services already appear to be going down this route.

Collaboration and data sharing between different services is also essential to the quality and breadth of data analysis. In the UK, collaboration between different public services has been made mandatory by underpinning legislation[3]. The requirement for fire and rescue services to develop four yearly integrated risk management plans under the Fire and Rescue Services Act, 2004 served to consolidate this framework for the analysis of information in light of effective planning. Due to the availability of big data through mandatory partnerships the analysis capabilities are immense. In 2014, the Fire Services Research and Training Trust awarded the National Fire Chief's Council a grant of £57,460 for a project called 'Sustained Action for Elderly Risk (SAfER)'. This project had sought to use a subset of fields from the National Health Service's Exeter System (also known as the NHAIS) to create a list including the address, year of birth and gender of each individual over 65 who is registered with a General Practitioner (GP) in England and Wales. This transfer of data has been intended to ensure that Fire and Rescue Services target preventative resources more effectively, at a time where the ageing demographic means fire deaths and injuries will increase significantly for the first time in 30 years (Cornford and Levin, 2017: p - iii).

In an interview with Head of Insight Analytics for the London Fire Brigade, Apollo Gerolymbos, in 2018, it was mentioned that such initiatives are only the tip of the iceberg. With reliable data, fire services are able to analyse almost anything. They can record and maintain lists of hoarders and allocate a risk rating that will indicate the extent of fire-loading that firefighters are likely to be confronted with in a fire at these premises. This will assist in determining the likely fire risk associated to the property which will in turn govern the choice of operational tactics as either defensive or offensive as well as the weight of response. They can record special hazards such as oxygen cylinders that are likely to be present on the property of individuals with respiratory problems or identify individuals who are unable to self-evacuate; they can identify previous fire

[3] Crime and Disorder Act, 1998; Civil Contingencies Act, 2004, Police and Crime Act, 2017.

incidents at specific properties and their cause; allocate resources to fit smoke alarms in the properties of high-risk individuals; ascertain a variation of risk information relating to specific types of properties and so on. Fire services acquire a plethora of information for commercial properties where specific risks are pertinent to fire service operations such a storage of hazardous chemicals, fire safety systems locations, access points etc. – and all of this can be available on the crew's Mobile Data Terminals[4] (MDTs) on the fire engines. Additionally, fire services are able to identify things such as patterns of firefighter injuries at different types of incidents, locations where assaults on firefighters have, and are likely to take place, as well as any other risk that will ensure that the incident can be dealt with more efficiently, effectively and more safely.

With regard to firefighter training for example, data analysis can establish the need for training against specific criteria on a firefighter-by-firefighter basis. As there is a requirement for firefighters to maintain competence, this would mean that a prioritisation system to train and retrain London Fire Brigade's 5000 firefighters would be established. As so much information is held electronically, the LFB does not only know when each firefighter needs to refresh their training in-line with the interval threshold, but they can substitute the frequency if the firefighter has attended many operational incidents within that period. As such, firefighters that have attended fewer incidents over a two-year period will be higher in priority to refresh their training than firefighters that are attending incidents more regularly. All such objectives are qualified through the analysis of decades of incident data in the region of multiple millions cumulatively.

---

[4] These are on-board mobile computers provided on fire engines that provide the crews with vital information relating the incident they are attending.

Another project named 'City Accelerator' undertaken in Louisville, Kentucky launched in 2014, was working on more than 50 projects under a multi-agency approach to address urban challenges through a new collaboration and data innovation process. By merging fire incidences and community service provision data, a geographic information system was created that helped the Louisville/Jefferson County Information Consortium determine that many fires occurred in neighbourhoods with high numbers of people served by the city's Community Services Department. The city also learned that thousands of abandoned homes and buildings presented a significant fire risk to neighbouring homeowners and community assets (Fox, 2016). This ultimately led to a series of initiatives to reduce the likelihood and impact of fire in abandoned properties and fire hot-spot areas.

In addition to this, Bernard Marr of Forbes wrote an article (2017) titled 'Is Big Data Analytics the Secret to Successful Firefighting?'. For this article, Marr interviewed senior firefighter with the City of Amsterdam fire department, Bart Van Leeuwen. Van Leeuwen argued that over the years he has come across several cases where firefighters were sent into blazes where it was later found that they had no business going into – but they didn't know that because they didn't have the data available to them. He stresses his concern about the potential of attending an incident where something might happen to him or his crew because they did not have adequate information for mitigating or controlling otherwise foreseeable risks. For a number of years, Van Leeuwen has championed the use of 'linked data' in firefighting. This concept is not just in relation to fire departments collecting as much data as possible through available information sources, but also data gathering from sensors attached to firefighting tools. He refers to this method as 'Predictive Modelling' which can be used to build up a picture of the risk profile of the area where the incident is taking place. Data taken from fire hose sensors and personal protective equipment can be used to build models to assess risk. Other information alike would determine how much water crews are using, how full their water tanks are, how much diesel fuel is left inside the fire engine, and the gas composition of the air at an incident. The basis for this process is to analyse all of this data and derive more effective models for getting information to the firefighters who need it.

The above are representative examples of how important data analysis can be for a fire service and highlight the extensive capabilities that are available. However, consideration should be given to how much data is enough data, what data should be acquired and when, and how it should be analysed. If these are not fully understood then the quality of that information is likely to be poor. If the quality of information is poor then the results and outcomes of the data analysis will also likely be poor (Quinn & Roden, 2014). It is understood that, not all fire services will have the capability to undertake effective data analysis that will be of statistical value to their organisation. This is either because they lack funding, resources or expertise, or perhaps a combination of all the above. It might also be the case that the data set is not big enough for there to be any statistical value to it. This would particularly be the case in smaller rural fire services, though little can be done to address the latter point and essentially data analysis in these cases may not be necessary. However, it is believed that despite big data analysis becoming the 'buzz word' of the last decade, it is still not fully understood and evidently not applied universally.

Understanding the data selected for analysis is crucial. If fire services do not understand the extent and types of risk that they are presented with then it is unlikely that they will be able to address these effectively. Though not all fire services are mandated to collect and analyse data that will be used to subsequently drive organisational objectives, there is a strong argument in its favour. Though some fire services might intuitively understand the intricate nature of emergency incidents that they attend, this knowledge is based on assumptions rather than accurate metrics acquired through reliable processes of data analytics.

In his article 'The Fire Service Future will be Data Driven' (2016), Rick Markley argues that sometimes the data that is derived from analysis, such as the most problematic fire locations, is indeed not news to experienced personnel within the fire service. However, he argues that *'it takes hard evidence to prove to those outside [the fire service] what all of us [firefighters] already know'.* Big data analysis provides reasonable justifications for fire services to be in the position to maximise their ability to tailor their services to meet objectives for protecting life and property. Most importantly it can assist them in understanding why they do what they do and direct them on how they go about doing it.

However, there is an antithesis and scepticism around the true intentions of the use of data analytics to improve fire service performance through effective resource allocation. It is suggested that data driven studies might sometimes be a political tactic for pushing through initiatives policy makers are in favour of (Ambinder, 2010). Perhaps one of the most well-known cases is around the RAND study that took place in New York City in the 1970s. The RAND Corporation was commissioned to conduct an in-depth analysis of fire department operations in New York City in the hope of combating the growing fire epidemic from the economic hardships the city was facing (Quinn and Roden, 2014). Joe Flood wrote a book titled 'The Fires' in 2010, that stated that in 1971, Mayor John Lindsay asked the Fire Department of New York's (FDNY) chief of department, John O'Hagan, for a few million dollars in savings to help close a budget deficit. O'Hagan turned to New York City-RAND Institute and the idea was to use data-based analysis to make recommendations to the FDNY to make the department more efficient and effective.

The project introduced a metric of performance-response times and computer models replicated when, where, and how often fires broke out in the city, and then predicted how quickly the fire department could respond to them. By showing which areas received faster and slower responses, RAND determined which fire stations could be closed with the least impact (Flood, 2010). During this period, there were areas of New York that were literally ravaged by fire. The Rand Report described that some fire stations in high incidence areas were responding to more than 6000 fires a year and more than 20 times a night on many nights. This extent of fire incidents was severely draining the city's resources (Blum, 1972 p - 4) and, hence, the prerogative for initiating new and effective solutions for dealing with this issue. However, it transpires that the RAND analysis suffered from several fatal flaws, which led to the recommended closure of several of the busiest fire stations in the most badly fire-ravaged parts of the city. In the synopsis of his book, Flood (2010) describes the situation as follows: *"Between 1968 and 1978, in New York City, a series of fires gutted the South Bronx, the Lower East Side, Harlem and Brooklyn, displacing more than six-hundred thousand people and killing thousands. Firefighters called it 'The Wear Years'. Conventional wisdom would blame arson, but these fires were the result of something altogether different: the systematic withdrawal of fire protection from the city's poorest neighbourhoods – a technocratic, money-saving decision with catastrophic consequences that would, astonishingly only set the template for how a modern city would function".* There was a strong evidential basis for the case of New York City and how flawed statistics can impact on public and firefighter safety with disastrous consequences.

Similarly, there was much contention and public outrage around a decision made to close ten London Fire Brigade stations in 2014. The London Fire and Emergency Planning Authority presented data that indicated that these closures would have no significant impact on response times and fire cover. These closures formed part of a cost-saving exercise of £45 million over a consecutive two-year period and were included in the Fifth London Safety Plan. The chairman of LFEPA confidently stated that *"Londoners will continue to receive one of the fastest emergency response times in the world from the London Fire Brigade. If you dial 999 and need a fire engine, we still aim to have one with you within six minutes and a second, if needed, within eight. The Brigade is faced with significant budget cuts which mean that changes to the service are inevitable, and we are able to make those changes without compulsory redundancies. The firefighters based at the stations closing will now transfer to other stations and continue the excellent work they do to prevent fires, which is vital in changing the behaviours that start fires in the first place".*

The press release also identifies that the Brigade's figures show that the number of fires in London had fallen by fifty percent in the last ten years between 2007 and 2017, and more recent figures showed that they continued to fall at the same rate in 2018 (London Fire Brigade, 2018). Conversely, the fire brigade union and mass media present a different story. An article in the Guardian written by serving firefighter and FBU chair for West Midlands Fire Service, Andrew Scattergood in (2017) argued that

> *"Since 2010, more than 10,000 firefighters have been axed, dozens of fire stations have closed, fire engines have been scrapped and levels of emergency rescue equipment has been slashed. In London, 10 fire stations have been closed, 27 fire engines axed, and more than 600 firefighter posts*

have been cut. Every year, response times are increasing and 2015-2016 saw a 15% rise in fire deaths compared with the year before. The fire service is finding it more and more difficult to respond to emergency calls efficiently and effectively. These cuts were instigated on the back of a global financial crisis that firefighters had no hand in creating. As a firefighter, I can say with authority that the public are not as safe as they were a decade ago. The increase in response times and fire deaths proves this. How is it that in the sixth richest country in the world emergency services are going backwards, damaging the safety of citizens?".

These competing views create a level of unreliability in not only the facts and figures (in terms of the raw data) and how they are interpreted, but also how these are utilised to satisfy specific objectives. However, the LFB Quarterly Performance report for Quarter 1 - 2018/19 formally presents the following data:

- The number of fires in the home (dwelling fires) continues to decline (5,584) over the longer term.

- The number of injuries from fires continues to decline (959) when compared to data over five years.

- A total of 82,848 home fire safety visits (HFSV) have been conducted and is well in excess of the target of 74,000 for 2018/19.

- The speed of attendance to incidents by both first and second fire pumping appliances continues to remain well within the performance targets (6 minutes and 8 minutes, on average, respectively). On average, a first appliance arrived at an incident in 5 minutes 11 seconds and a second appliance in 6 minutes 31 seconds. The first appliance arrived within 10 minutes on over 96 per cent of occasions.

- However, the **all-fire deaths** (10-year average) remains above target (52), along with accidental fire deaths in the home (10-year average) (39). In the 12 months to the end of Q1 2018/19, there were a total of 44 fire fatalities, compared to 109 in the same period for the previous 12 months.

- The Brigade is also exceeding its sickness targets for all staff groups, with sickness above target levels (i.e. more staff reporting sick) for Operational staff (5.47 per cent), Fire and Rescue staff (3.88 per cent), and Control staff (7.66 per cent).

In an age of austerity, where fire services are asked to do more with less, justification must be given to cost saving objectives when these are likely to impact on performance targets. Data analysis can support such exercises. However, the contention lies in respect of whether such evidence can be created or doctored in such a way so as to satisfy pre-arranged political goals.

De Haan and Icove in their book 'Kirk's Fire Investigation' (2014: pp 10 -11) argued that the major issue in identifying and quantifying the fire problem is the failure to obtain accurate incident data. They further state that this problem is not limited to any single country. In the United States, as a result of the convoluted and complex systems of recording fire data, the Consumer Product Safety Commission estimated that fewer than 5% of structure fires are reported to the authorities. Additionally, it is argued that many state fire marshals do not report all their fire incident data to the National Fire Incident Reporting System (NFIRS) due to shortages in staff.

Furthermore, the state of California, that represents more than 10% of the U.S. population is not represented on the NFIRS for several years.

With regard to the U.K., findings from a 2004/5 Survey of English Housing on all outbreaks of fire experienced by households in England suggested that the fire and rescue service attended approximately one fifth of all domestic fires. The explanation for this was that many fires recorded in this survey were only minor allowing residents to take action themselves without summoning the emergency services. However, it appears that regardless of whether all fires in the UK are in fact reported, those that are, appear to provide sufficient data to create statistically meaningful interpretations and subsequent initiatives to address them by UK fire and rescue services. The declining numbers of fires, firefighter fatalities as well as the generally low number of fire related civilian fatalities overall would indicate that there is value in the current data for UK fire services that collect and analyse them as part of their strategic management profile that allows them to be in a position to effectively address overarching goals for life safety and property protection.

Data and its analysis can greatly assist fire services in developing accurate objectives around overall service delivery in matters such as fire prevention, the allocation of funds, fire service operational tactics, equipment, resourcing and training (FEMA, 2021). It would indeed be irregular for any fire service to not utilise such a system, especially in times of financial crisis where there is greater impetus to justify expenses and work in a cost-efficient manner. Contrasting evidence, however, indicates that data analysis is not always objective or accurate. If there is no foundation on the type and quality of data, then the results will not be reliable. Furthermore, there is some evidence to indicate that data analysis can be undertaken in such a way so as to promote specific political goals rather than organisational objectives. Nonetheless, data analytics is an extremely valuable tool if used correctly and has many benefits if its focus is set in the right direction without overbearing political influences.

Fire Service Organisational Culture

There are numerous definitions of culture, some of which are more generic and those that are more specific to the intrinsic characteristics of certain groups, such as firefighters. One generic example has been given by the Multi-Agency Serious Accident Investigation Team established to investigate the injury of a Fresno Fire Department Captain who fell through the roof of a burning structure during a vertical ventilation operation[5]. The Cortland Fire Incident Report that was published as a result of the investigation (2015: p 53) states that culture is, *"the sum of attitudes, customs, and beliefs that distinguishes one group of people from another. Culture is transmitted, through language, material objects, ritual, institutions, and art, from one generation to the next"*. In society, individuals, as well as groups, can identify with (and be identified by) one or more cultural profiles. In a collective sense, a simple example would be how people identify with the dominant cultural values of their country, values that give them unique identity. These values are referred to as heritage culture or mainstream culture (Yampolsky, Amiot and de la Sablonniere, 2013).

---

[5] The American fire service traditionally utilises crews to ventilate burning structures from the roof. Crews gain access to the roofs of burning structures and with the use of various tools open holes so that smoke, heat and fire can be ventilated into open air. The premise is that the smoke layer in the property with rise, giving better visibility, releasing pressure and decreasing heat which will substantially improve both survivability of victims as well as assist firefighters locate and deal with the fire. This task has a number of inherent risks on one which is risky with regard to the structural integrity of the that is being heated from the fire below.

Language is a characteristic example of heritage culture as are traditions, rituals and other specific behaviours, though specific behaviours can be influenced by mainstream culture as well. These cultural traits give people distinguishable ethnic identity i.e., Greek, French, English, Italian and so on. Within dominant culture, there also exist smaller subsets. These can be formed between groups of people that share similar values, purpose or beliefs. Religion is a prominent example as this can be reflective of dominant culture. For example, Christianity reflects a broad religious category that is split into subsets such as Orthodoxy, Catholicism, Protestantism and so on. These religious affiliations will be represented more widely by geographical indicators i.e. Orthodox within the Balkans, Russia etc., Catholicism and Protestantism in Europe, whereas the Middle East is more widely represented by Islamic factions (Park, 2004: p-6). Each one of these religious groups will possess a set of cultural values and identifiable characteristics.

Similarly, people within their daily work environments will develop a cultural profile that will be informed by the intricacies specific to their type of work. Organisational culture can be understood as a subset of dominant culture. However, organisational culture also possesses a hierarchy – with a dominant culture and subsets. It is suggested that cultural hierarchy (between dominant culture and subcultures) can be understood better when evaluated within a specific context. Katib (1996: pp 5-6) argues that most large organisations have a dominant culture and multiple subcultures. Many of these sub-cultural groups will have characteristics of dominant culture (especially mainstream culture) but will be represented by unique features that will be specific to that trade. In a fire service context, this could be understood as the organisation being represented by a broad set of values, or rules, implemented though policies and code of conduct, thus establishing the dominant cultural profile[6], whereas at shift, rank and fire station level, and even by specialisation, there are a number of sub-cultures with their own characteristics and values that give them identity and structure.

In a joint report published by the International Association of Fire Chiefs and the United States Fire Administration (2015: p-3) regarding a 'Study of Behavioural Motivation on Reduction of Risk-Taking Behaviours in the Fire and Emergency Service', it is stated that *'[fire service] culture has not been concisely defined, but literature suggests both that it exists as a stand-alone concept and that it has unique characteristics that are uncommon to non-uniformed professions'*. The Cortland Fire Report (2015: p-53) attempts a more precise definition in respect of fire services. It states that, *"Cultural values and attitudes influence many factors within a fire department…. Organisational culture is very difficult to define and measure, although certain characteristics tend to be easily recognizable. Organisational culture is the product of a particular combination of influences including leadership, traditions, experiences, peer pressure, and discipline"*.

Schein (2004: p 17) asserts that culture is *'a pattern of shared basic assumptions that were learned by a group as it solved its problems of external adaptation and internal integration, that have worked well enough to be considered valid and, therefore, to be taught to new members as the correct way to perceive, think, and feel in relation to those problems'*. This process is fundamental to how fire services operate. The work of a firefighter is technical and requires skill to be undertaken effectively. A skilful firefighter will acquire status and acceptance within their workplace and, therefore, this attribute

[6] Organisation-wide code of conduct will define the core values and acceptable behaviours that will ultimately shape the dominant organisational culture as this sets an expectation that all employees within the organisation must conform to.

(technical ability) can be seen as holding significant value within fire service culture (Turnell-Read and Parker, 2008: p - 130). Operational experience is another attribute that holds significant weight within the context of fire service culture and asserts individual credibility and status within a group (Shaw, 2011).

Conversely, Thiel (2013: p-21) argues that *'fire service culture typically emphasizes aggressiveness, an action and the ability to overcome obstacles in the course of mission performance, while at the same time upholding a hierarchical organisational structure for fire-ground decision-making'*. Additionally, in his thesis, 'the Significance of Fire Service Culture as an Impediment to Effective Leadership in the Homeland Security Environment', Cox (2012: p–31), discusses the intricate elements that characterise fire service culture. Amongst other attributes, he refers to the hegemonic masculine culture, the dynamic of rank and file, the value of story-telling, symbols of pride and identity, the establishment of a safety culture and implications of risk-taking behaviours that arise from complacency as a result of conditioning due the frequent exposure of firefighters to risks that are most often dealt-with without incident.

In relation to masculine culture, Brunsden, Hill and Maguire (2014: p - 6) in their paper 'Putting Fire and Rescue Service Stress Management into Context: A United Kingdom (UK) Informed Perspective', argue that despite efforts to diversify fire and rescue services in the UK by promoting numerous campaigns to attract more women and minority ethnic individuals, fire and rescue service operational personnel remains dominated by white males. Though they argue that there are positives in respect of the solidarity established through notions of a brotherhood or fraternity amongst male firefighters, these characteristics may serve to deter prospective female applicants. It is found that this gender imbalance creates a particular form of organisational culture — one that can detrimentally impact on stress management initiatives. It is argued, that as males are more reluctant to publicly share their emotions, this serves to generate and maintain barriers to help-seeking. The almost exclusively male population, combined with the perception of fire and rescue service's work as heroic, can further promote a macho organisational culture which has been referred to as one possessing the three T's "testosterone, tattoos and taut-biceps" (Beale, 2003, p - 29).

These matters have come to the forefront of the media attention recently as a result of a Her Majesty's Inspectorate of Constabulary & Fire and Rescue Services (HMICFRS) report 'Values and culture in fire and rescue services' (2023) which found that found that examples of racist, homophobic and misogynistic behaviour in a quarter of fire and rescue services in England, with such behaviour often excused as banter. These findings were further reinforced by an independent cultural review of the London Fire Brigade undertaken by the former Chief Crown Prosecutor for north-west England, Nazir Afral OBE. The review was commissioned by the London Fire Commissioner in response to the suicide of firefighter Jaden Matthew Francois-Esprit, who took his own life in August 2020. The review ultimately concluded that London Fire Brigade as an organisation is institutional racist and misogynist (Winchester, 2022).

With regard to fire service culture, Bill Gough discusses a far simpler definition that effectively encapsulates all of the above definitions into one phrase; fire service culture can be understood as *'the way we do things around here'*. He argues that there are many factors that influence behaviour, but that focus should be placed on when is 'the way we do things around here' concealing cultural behaviours based on 'we've always done it in this way' rather than through formal and tested methods (Gough, B. cited in Grimwood, 2017: p - 443). The *'way we do things around here'* isn't necessarily bad if it is supported with an adequate balance of experience and science (Hartin, 2013). If, however, the sake of doing things in a particular way is simply based on experience and tradition alone, then risks to both firefighters and the public are likely to be higher as errors are more likely to occur. Gough further negotiates the relationship between habit and culture and how the 'moment of choice' can invariably be influenced by routine practices based on habit. The 'moment of choice' refers to the critical moment where informed decisions have to be made to address an incident to bring it to a safe resolution. He argues that 'skill' and 'expertise' are, in fact, effectively well practiced habits that enable people to work on auto-pilot.

James Reason (1997) further asserts that 'skill' represents routines which are highly practiced tasks that people are very good at 'most of the time'. However, Reason points out that strong habit intrusions are among the most common of all error forms which are most likely to occur during the performance of highly automated tasks. This is described as one of the downsides of habit as people get so used to doing things on auto-pilot that they fail to identify subtle changes that occur in their work environment when things are almost but not quite the same. The Cortland Report (2015: p - 53) defines this process as 'the Normalisation of Risk' which occurs when an organisation slowly accepts dangerous behaviour when immediate severe consequences do not develop following the behaviour. It was also identified that field level operating procedures in the Fresno Fire Department relied more heavily on tradition and department culture, many times in direct opposition to department policy.

In order to counter such incidences, one would expect that the systematisation of functions through the implementation of Standard Operating Procedures (SOPs) would serve as an ideal control measure. However, as argued by Reason (2000) *"the basic premise in the system approach is that humans are fallible and errors are to be expected, even in the best organisations. Errors are seen as consequences rather than causes, having their origins not so much in the perversity of human nature as in "upstream" systemic factors. Counter-measures are based on the assumption that though we cannot change the human condition, we can change the conditions under which humans work".* Conversely, Gough argues that habitual working has the tendency to overlook SOPs meaning that how work is planned is often not how work is always done.

Reason (1997) describes this error type as a 'violation' and suggests that attempts to control behaviour through the creations of SOPs without considering the wider cultural implications and constraints might have an adverse effect by increasing the likelihood and opportunity of such violations. It is stated that when behaviours are controlled by decision rules such as 'must' and 'under no circumstances', procedures will be routinely violated or when an incident does not fit the SOP model then an exceptional 'work-around' or adaption is used. The dilemma this creates at the 'moment of choice' is when to comply and when to violate and this is where the link between habit and culture is possibly at its strongest, when routine violation becomes the alternative route, ingrained within the 'the way we do things around here'. When these habitual nuances are not realigned in the workplace, habitual behaviours become routine violations and firefighters get hurt (Gough cited in Grimwood, 2017: p – 444). In support of this view, the Cortland Fire Incident Report (2015: p – 54) identified that:

> *"When dealing with department policies and procedures, it is quite apparent that the membership [of Fresno fire Department] at all levels is picking and choosing which ones they are willing to follow, including when the policy states, you "shall". The culture of the leadership at all levels has resulted in the lack of enforcement and*

accountability. Department members have stated during interviews that, "providing more policies and sending them out with no demonstrated effective leadership and or accountability will not fix the problem". Many of the policies the department utilizes are well written but not followed-up on to ensure a change of behaviour".

This in itself demonstrates the problematic relationship between leadership and culture as identified by Reason (1997). On the one hand, habit will have a direct effect on the enactment of operational procedures at the moment of choice whilst, on the other hand, it appears that the development of policies and procedures does not often consider their applicability at a practical level and are perhaps not enforced appropriately. Developing standards is a useful tool to ensure that work will be undertaken against specific guidelines that will ensure mission objectives, such as safety and speed. If there is no process to measure the effectiveness of these procedures, then they simply exist as a point of reference that firefighters can delve into as part of their toolbox to support incident management.

In his article, 'Human Factors and the Fire Service', Kerwood (2009), discusses the errors that led to the fatalities of 14 firefighters in the Storm King Mountain Fire in Colorado (USA) in 1994. He reveals that one of the Storm King Mountain firefighters, in an interview, stated that the rules that were broken the day of the tragic fire were 'too many to count'. It was also identified that this negligent attitude toward personal safety was illustrated repeatedly during the interviews with the surviving firefighters of the Storm King Mountain fire crew. Further to this, another surviving firefighter explicitly noted that it was well within the job of the wildland firefighters, based on their knowledge, skills and abilities, to violate policies and procedures if it meant getting the job done. Though human errors are identified as one of the leading causes of firefighter Line of Duty Deaths or serious injuries (Moore-Merrell et al, 2006), the National Fallen Firefighters Foundation (NFFF) has asserted that the culture of the fire and emergency service is a major contributor to the fatal trend in firefighter health and safety (Siarnicki, 2010).

When analysing the structures of some of the more highly regarded fire services around the world, despite their robust systems of service delivery and rhetoric around safety, the issue of fire service culture, to a greater or lesser extent, appears to be recurring as a key contributing factor in firefighter injuries and fatalities on a global scale. This would mean that despite efforts to ensure the implementation and maintenance of a strong safety culture, its effectiveness is often dependent on a number of complex and interrelated factors. These factors will influence organisational culture across the entire hierarchy (dominant and subcultures) and certain characteristics will unequivocally be influenced by their respective dominant cultures (heritage and mainstream). This is what makes the issue of fire service culture so complex to analyse and overcome if not considered within each fire service's (or nation's) specific context.

Indeed, there are many commonalities that would define fire service culture on a global scale; the fact that it is a predominantly male occupation; there is significant value placed on experience and technical competency; there is a tendency around storytelling and a reluctance around openness and sharing of feelings amongst male firefighters etc. Fire service organisations implement a rank-and-file system and follow a hierarchical structure with defined roles and responsibilities. Aside from these generic commonalities, there are marked differences that distinguish fire services from one another and the extent to which culture has a negative impact on elements such as safety and service delivery are arguably highly contextual.

Fire services in Northern Europe are generally more organised and systems-orientated than those in Southern Europe. This is what appears to prevail from the research questionnaires analysed later in This book. There is also plenty of evidence that has been identified through the investigative process in relevant literature, documents, articles, policies and standards as well as through interviews with fire service representatives from both Southern and North European countries respectively. North European fire services tend to have processes and procedures that define their management and operational structures, and these, especially in more recent years, have acquired a business model approach under the auspices of New Public Management (Carvalho, Fernandes, Lambert and Lapseley, 2006: pp 168 - 170). The increasing utilisation of data analysis also supports investment and budgeting initiatives, as these tend to identify problems, or needs, through valid metrics instead of reactive supply and demand type structures of more administrative models. Generally speaking, the ethos around these processes is also prevalent within dominant culture of North European countries.

Social and Political History and its Role in Defining Culture

In order to understand the stark differences in culture between fire services in Northern and Southern Europe, one must look into the social and political histories of the countries in question. It will be argued that events from the past of the south European countries in this study will ultimately shape their dominant cultural characteristics today. These cultural traits are most resistant to change and often those that broadly define their peoples by ethnic group or geographical location. These are also transferable across a number of dimensions including social networks, the family unit and the work environment. There are many populist views, or stereotypes, around the organisation and efficiency of North Europeans.

A study by Associate Professor and Director of the Danish Social Capital Project, Gert Tinggaard Svendsen (2003: p – 15) found that in the EU, 'Northern EU' member states (Finland, Denmark, Sweden, Netherlands, UK, Germany, Ireland) generally hold less corruption, more trust and higher Gross Domestic Product per capita compared to 'South EU' member states. Portugal, Italy, Greece and Spain, also referred to contemporarily as the 'PIGS', a name that was acquired during the period of financial crisis circa 2008 to abbreviate those countries most hardly hit by the economic downfall (Lewis-Beck & Nadeau, 2012: p - 472), are not generally characterised as organised, hard-working countries with meritocratic systems and processes in place to enable efficiency. Contrarily, and despite boasting a unique way of life, the PIGS are more often than not culturally represented by significant levels of corruption, lawlessness and disorganisation (Zamora-Kappo & Coller, 2014).

An article regarding corruption in Southern Europe by the Legatum Institute in November 2016 stated that *"there has always been corruption in Southern Europe....so much so that the abuse of a system is becoming the system itself"*, and there is extensive research on the problems of corruption and clientelist relationships that are so widely spread throughout these countries. According to another study by Lacko (1998) on shadow economies estimated that two south European countries, Greece and Italy, had an underground economy almost one third as large as the officially measured GNP, followed by Spain, Portugal with a shadow economy between 20-24 % of official GNP (Katsios, 2006: pp 66 - 67). Corruption is part of everyday life, and many systems and processes operate with corruption and clientelism at their core. People placed in positions of power especially in the public sector are often appointed as a result of their political affiliation (Danopoulos, 2015: p – 133).

This demonstrates that for appointment into senior roles, emphasis is often placed on political affiliation rather than objective and meritocratic criteria. This creates a series of problems that will be discussed later in the case study analysis. However, what one must consider is that where such deeply rooted and widely spread corruption exists, this will also affect the broader culture of public sector organisations, and this is the fundamental difference between Northern Europe and Southern Europe. In the former case, there is mostly a culture around rules and regulations that is generally adhered to creating robust and auditable structures. These structures set the organisational foundations of fire services through transparent and meritocratic processes. Conversely, in Southern Europe there is a generalised lack of formal systems and processes. Concepts such as transparency and meritocracy do not hold great value at organisational level. Clientelist relationships and political affiliation appear to be the dominant features that influence key aspects of the organisation that often serve to severely hamper any progress or development.

It is interesting that in Northern Europe there is plenty of literature around fire services and how they work, there is availability of numerous firefighter fatality case studies, training articles, manuals and so on. This would indicate that there is generally a willingness to examine the core of the organisation and understand how it does what it does and how well it does it. This has to do with the perception of this profession as a vocation rather than just a service and the drive to promote excellence and professional standards. This can only be achieved by exposing weaknesses and then addressing them in light of improvement and development. Though some countries in Southern Europe are gradually participating in international forums around fire service development, they are still greatly under-represented. These organisations appear to be protective and institutionally isolated, what Argyris & Schon (1978) refer to as 'organisations with defensive routines', a characteristic that will ultimately serve to hinder learning.

Consequently, it appears that the relationship between corruption and culture is very complex and contextual, and corruption manifests itself in different ways that influences the extent and impact it has on organisational culture. Even where similar mechanisms of corruption are identified and are at play, the extent of its influence on integral functions between organisations can in fact be very different. Therefore, it is this complex relationship between corruption and culture that This book aims to address, especially where similar mechanisms of corruption can be identified. If these questions can be answered, it is then pertinent to determine whether, in such cases, one must simply accept that where significant corruption is ingrained within organisational culture, the fire service and the community will invariably be at greater risk. If, however, this is not an acceptable outcome then it must be determined how one can effect change.

Understanding Greek Culture and its Impact on the Fire Service

Greece has had an interesting yet troubled history with very few periods of social, political and financial stability, especially over the past 200 years. One of the most important periods of Greek history that has contributed to some of its key cultural characteristics around corruption and general lawlessness stems from the Ottoman Empire between 1453 to 1821. When Greece was conquered by the Ottoman Turks in 1453, this led to a 400-year occupation where ethnic identity and national borders became undefined and intermixed. During this period Greek speakers were spread across Albania, Macedonia, parts of the Balkans, southern Russia, Ukraine, Asia Minor, the region of Pontus, Eastern Anatolia, Georgia, the South Caucasus, Egypt etc. (Korma, 2017). Many also sought refuge in Italy, in cities like Venice, Florence, and Rome (as well as other European cities) and bringing with them a cultural heritage that contributed greatly to the Renaissance (Kouras, 2018). Within what are contemporarily the borders of Greece and particularly the region of Macedonia there lived numerous other communities of Vlachs, Slavs, Bulgarians, Romanians, Epirotes, Christian Turks, Muslims. etc that, however, identified as Greeks (Kalyvas, 2015: p - 45). Nationalities were intermixed but Hellenism was broadly held together by language and religion rather than geography even though it was not until the mid-twentieth century that the Greek language was officially consolidated into what it is today (Diatsentos, 2016: pp -215 – 216).

An important point to note, is that the Greek Orthodox Church and the Ecumenical Patriarch were appointed as the religious leader of Christian Orthodox communities in all regions under Ottoman control. However, it is suggested that this eventuality came about as early as 1466 through bribes offered to the sultan by aspirants of the patriarchal throne (Orthodox Church in America, 2024). This relationship inevitably led to the Orthodox church to fall victim of the institutionalized corruption of the Ottoman system of government. Thus, the combination of civil and religious power in the hands of the ecumenical patriarchate and the upper reaches of the hierarchy prompted furious competition for high office.

The Ottomans encouraged such behaviour, and it soon became the norm that, on every occasion when a new patriarch was elected, a huge 'peshkesh' (bribe), would be paid to the grand vizier, the sultan's chief minister (Delano et al., 2024). As such, the patriarchal office increasingly became the object of bribery across all levels of society, a form of corruption that is widely known as 'Rouspheti'. Clogg (2013: p-4) asserts that 'Rouspheti', which is the reciprocal dispensation of favours that has traditionally oiled the wheels of society, and *mesa*, the connections that are useful, and indeed indispensable in many aspects of daily life, were both reinforced during the period of Ottoman rule.

Clogg (2013: p-3) also describes the general attitude of the time and perhaps the need to foster such clientelist relationships enabled Greeks to acquire a reasonable standard of living, satisfying various wants and needs. He asserts that *"the capriciousness of Ottoman rule and the weakness of the idea of the rule of law helped to shape underlying values of Greek society and to determine attitudes to the state and to authority that have persisted into the present. One form of self-defence against such arbitrariness was to secure the protection of highly placed patrons who could mediate with those in positions of power and privilege. This was coupled with the distrustful attitude towards those outside the circle of extended family".* This asserted a greater level of individualism and is perhaps one of the underlying causes of the failure of Greeks to create strong community bonds and collective aspirations as will be seen throughout this section.

The war of independence in 1821 was founded on sentiments around the rich culture of ancient Greece. Greeks of the intellectual class living and studying abroad in Italy, Switzerland, France and Germany became aware of the reverence with which the language and culture of ancient Greece were beheld throughout Europe. This served to promote a sense of national pride amongst the Greeks, though this was initially only amongst the intellectual classes and mainly outside of Greece. Over the years, there were some unsuccessful initiatives to promote the idea to liberate Greece from the tyranny of the Ottoman Turks but the drive to restore the grandiosity of ancient Greece eventually came to fruition by three young Greeks, Emmanuil Xanthos, Nikolaos Skouphas, and Athanasios Tsakalov, lowly members of the Greek mercantile diaspora, who in 1814 in Odessa (now in Ukraine), the centre of a thriving Greek community, founded the Philikí Etaireía, or "Friendly Society."

On 25th March 1821, a revolt against the Ottoman Turks commenced. With the support of foreign forces, the Ottoman's were overthrown, and it is from this point that Greece was again recognised as an independent nation. However, following the war of independence, matters did no go as planned and significant infighting and political rivalries ensued that led to European intervention allocating a series of foreign King's to instil peace and oversee the establishment of a new government. This foreign oversight persisted in Greece intermittently until the 1970s. Until the early 1900s, Greece was still trying to establish itself as a nation but was overwhelmed by internal strife. Until the 21st Century Greece has experienced a number of critical events that would shake its social structure to the core.

Amongst others, some of the most notable are:

- The Macedonian Struggle (1893 – 1908).
- The National Schism (1910 – 1922) that separated the country into ardent royalists versus government supporters 'Venizelists'.
- The Balkan Wars (1912 – 1913).
- World War I (1914 – 1918).
- The Anatolian Disaster (1919 – 1922) that saw a fierce retaliation by the Turks to a Greek offensive in Asia Minor that led to thousands of fatalities and the progressive mandatory exchange of populations with an influx of 1.2 million Greeks returning to mainland Greece.
- The Great Depression (1930s).
- World War II (1939 – 1945).

- The Greek Civil War (1946 – 1949) that occurred between the Greek State military and the Democratic Army of Greece, the military branch of the Communist Party.
- The Military Junta (1967 – 1974), a military dictatorship characterised by right-wing cultural policies, restrictions on civil liberties, and the imprisonment, torture, and exile of political opponents.
- The Turkish Invasion of Northern Cyprus in 1974 that took advantage of the civil unrest in Greece.
- The entrance of the socialist government in 1981. Though this era came with a reasonable level of social stability it is under this government that most irreversible damage was done to further facilitate the clientelist relationships of the past with examples of outrageous levels of corruption.
- The Financial Crisis of 2008.

These historical events have had a tendency of consistently separating Greek society into factions (Greeks within the borders of independent Greece and Greeks outside the borders of independent Greece, peasants and intellectuals, nationalist Greeks and Ottoman Greeks, royalists and Venizelists, democrats and communists, communists, and dictators and so on). There was also a question around ethnic identity, who was in fact really Greek and what did 'being Greek' actually mean? Was it based on geographical factors; through language or religion; or did it perhaps have to do with heritage? These questions were at best challenging to answer after 400 years of Turkish influence and such a mix of ethnic communities across the Empire and specifically within Greece. However, it is believed that this persisting social divide would ultimately serve to promote longstanding instability and uncertainty in the emerging Greek state, even to this day. Additionally, where such turmoil existed within Greek society and across the entire political spectrum, the failure to acquire any financial equilibrium should not appear strange. This is evident throughout Greek history over the last 200 years where Greece has constantly required to default on financial deficit or seek assistance from Europe to pull through economically treacherous times.

However, even after its independence from Ottoman rule, Greece has managed to demonstrate a firm assimilation to the corrupt practices associated with clientelist relationships, bribery etc. of the Ottoman era. This characteristic appears to be deeply rooted in Greek culture and indeed one of the main features of Greek society contemporarily. Clogg (2013: pp 3-4) described this process after the emerging state in 1821. This description reigns true to this day. He states that, *"the need for new patrons continued into the new state and, once constitutional government has been established, parliamentary deputies became the natural focus for clientelist relations, which pervaded the whole of society. In return for their support at the hustings, voters expected from those whom they had voted to help them and their families to find jobs, preferably in the inflated state sector, the only secure source of employment in an undeveloped economy, and to intercede with a generally obstructive bureaucracy"*.

Moving forward to the 1980s, the socialist government under Andreas Papandreou (PASOK), indeed, exploited this trait to the fullest in the upcoming elections. In fact, it is argued that the 1980s was the decade that corruption became institutionalised in Greece. It stated that between the election of the socialist government in 1981 and 2010, scores of major corruption scandals (some count over 150) broke out in the press (Zoakos, 2010: pp 18 – 19). One of the principal themes associated with Papandreou's socialist government was the massive expansion of the public sector. It was identified by Karabelias (2013) that staffing in public organisations increased from 350,000 in the late 1970s to over 800,000 by 1989 (Avrantinis, 2013). It is also mentioned that in the last month of the pre-election period in 1989 alone, the socialist government appointed 180,000 employees into the public sector whilst receiving multiple loans from foreign lenders to support this initiative (Kathimerini, 2011; Manitakis, Marr 2012; Demetis, 2016).

It is not uncommon for many public servants, including firefighters, appointed into state organisation roles during the 1980s to openly state that they were employed with 'a tin of oil'. This, in its literal sense, meant that a family member had visited a political figure and gifted (bribed) them and promised them votes in the upcoming elections. The 'tin of oil' as a metaphor had many shapes and forms ranging from the exchange of livestock, food, beverages, other items of value and in many cases included outright bribery with money in return for 'favours'. The outcome of such transactions commonly saw the appointment of family members of the bribers into public sector organisations including the fire service. Though such accounts are often anecdotal, it is argued that by 1984 it was estimated that 89 % of all card-carrying members of PASOK had some professional affiliation with the public sector either through a permanent job, a temporary job, or a contract to do business with it (Kalyvas, 2015: p - 137).

It is also argued that through the introduction of a devastating legislative trifecta, the PASOK government undermined any possibility of meritocracy and accountability in the public sector. These legislative innovations reinforced several harmful tendencies of the Greek bureaucracy which included mismanagement, laxity, lack of accountability, worship of highly formalistic and rigid legalism, prevalence or everyday informal practices that led to petty corruption, absence of sanctions for those caught breaking the rules, and the abuse of discretionary decision-making. In addition, staff effectiveness was undermined as personnel evaluations required the signature and consent of those being evaluated and the process of employment into public sector roles changed dramatically after 1981 eliminating the competitive examination recruitment in exchange of an ad-hoc process founded on clientelist relationships and arbitrary criteria. The ensuing rise in partisanship and the collapse of professionalisation caused a marked decline in the quality of public administrations (Kalyvas, 2015: pp 136 – 137). It is noteworthy that a study undertaken in the 1980s identified that of the young Greeks (under 25 years of age) 64 per cent aspired to hold a job in the public sector (vs. 26 per cent in the other EEC countries), and only 10 per cent in the private sector (Dimitras, 1990: p – 105). This culture within and about the public sector is still prevalent today.

Where a system, such as that in the Greek public sector has such long standing roots around corrupt practices, these undoubtedly serve to discourage development, meritocracy, innovation and professionalism. Indeed, what are the incentives to strive for excellence when there is no obligation (or repercussions) to foster such practices? The Hellenic Fire Service has itself been involved in a number of scandals relating to the mismanagement of funds around public tenders, the ambiguous appointment of officers into senior roles outside of their promotion queue, thus superseding other officers earlier in the promotion pool (Lambropoulos, 2020). There is also likely to be a proliferation of corrupt practices at lower level that nonetheless contribute to the overall organisational culture.

Recruitment campaigns have also, in many cases, been photographic meaning that the recruitment criteria made it almost impossible for any other candidates to fulfil the objective criteria of the advertised roles. This is most characteristic for the recruitment process that aimed and making permanent 5500 seasonal firefighters with long service over a series of consecutive forest-fire seasons. For the first time, the fire service increased the age limit from 33 to 50 and gave numerous points (that were associated to an employment ranking system) to anyone who had worked as a seasonal firefighter in the past. For example, volunteers were granted 1000 points for every year they had served as a volunteer up to a maximum of five years. The age limit was also much lower for volunteer firefighters applying with a maximum age limit of 33. Conversely, seasonal firefighters received 2000 points for every year they served without a limit on the number of years. With most seasonal firefighters serving a minimum of five to eight years, there was little competition and through consecutive recruitment campaigns a significant number of the 5500 were made full-time firefighters.

The Impact of the Financial Crisis on Fire Services

It appears that the global financial crisis is having a wider impact on the quality-of-service delivery of many fire and rescue services that traditionally, and comparatively, have been industry leaders in promoting service excellence and organisational development. In the UK, the requirements placed on fire services under the Fire and Rescue Service Act in 2004 intended to promote a level of functional independence that was based on specific risk criteria pertinent to the Local Fire Authority areas. The implementation of Integrated Risk Management Plans ultimately served to create a balance between effectiveness and efficiency against cost-benefit analyses and identified localised risk criteria.

As a result, many fire services have had to adapt traditional methods of service delivery, as a standardised national approach under this new localised system was no longer feasible. In the first instance, it is clearly identified that National Operational Guidance is not intended to dictate or prescribe how competence should be achieved or how training and development programmes must be delivered, which is a matter for each individual UK FRS to establish based on their circumstances. It is expected that the guidance can be adopted voluntarily across the UK and is available for individual fire and rescue services to consider its implications on their own policies, procedures and training (Department for Communities and Local Government, 2012: p -3).

It is further stated that some of the guidance may not be relevant to all fire and rescue services as this will depend on the range and scope of services the fire authority has agreed will be provided, usually through its integrated risk management plan (National Fire Chief's Council, 2024)). The new structures that fire services have adopted in order to balance their effectiveness against cost-saving initiatives means that traditional methods are not always suitable or applicable, especially for smaller crew sizes, a format which appears to be favoured by many fire services contemporarily across the UK. Though the intention of the Fire and Rescue Services Act, 2004 was stated to positively address matters such as de-centralisation, the shift from emergency response to prevention and greater collaboration between relevant stakeholders (LGA, 2017: p – 15), it appears that there was also a significant down-side to this new structure.

The Fire Brigade Union has consistently highlighted their concerns in relation to the negative impact that these changes will have on service delivery, creating increased risks to both the public and firefighters. In a press release in 2004 the FBU argued that *"fire authorities are responsible for the level of local services but with no national standards to underpin them; It leaves the absurd position of having national nutritional standards for school meals but no national standards for fire cover (how many firefighters and how quickly they respond to incidents)"*. The FBU also warned that *"regional emergency fire controls are planned at massive expense which will be more vulnerable to breakdown and will damage the ability to deal with major terrorist incidents"*.

In 2016, the Chief Fire Officer's Association published a report 'Fighting Fires or Firefighting: The impact of austerity on English Fire and Rescue Services'. In this report, they highlighted their concerns around the pressures to reduce budgets and the impact that this might have on overall service delivery. In a survey to identify how Chief Fire Officers would continue to deliver a balanced budget, it was found that:

- A further 4,050 frontline firefighter posts will be cut during 2013-15.
- Almost 1,150 additional jobs will be cut from back-office support functions.
- 150 fire appliances will be cut.
- 60 fire stations will close.
- Almost 40 other fire stations will provide a reduced level of operational cover.

This will mean that over the Spending Review period, fire and rescue services will have 22% less frontline staff to deliver the high impact outcomes and that, as a consequence, it will be their ability to prevent emergency incidents, to reduce demand on other public services and to make a positive contribution to economic recovery which will be severely compromised (CFOA, 2016: p – 16). Furthermore, in the summer of 2017, HMICFRS was established to assess and report on the efficiency, effectiveness and leadership of Fire and Rescue Services in England. HMICFRS are an independent body appointed by the Crown that aims to provide authoritative information to allow the public to compare the performance of their fire & rescue service – against others. The evidence provided is used to drive improvements in the services they provide to the public. All 45 FRS in England were inspected in three consecutive tranches of 15 FRS, beginning in the summer 2018 and concluding in Autumn 2019.

The reports identified that although most FRS are good at responding to emergencies, there is too much variation in how well the public is protected, how quickly emergencies are responded to and how well services look after their staff. The findings from tranche 2 also warned that more than a decade of localism created as a result of requirements made under the Fire and Rescue Services Act 2004 had led to marked differences between services, for example, in how they have determined their response standards and record them; how they identify and mitigate risk; and how they define and audit high-risk premises. In more detail the report states that, *"we are pleased that fire and rescue services show real strengths in training for and responding to emergencies – this work undoubtedly saves many lives. However, it is concerning that there is too much variation in how fire and rescue services operate, resulting in a postcode lottery in the standards of service the public receives. We were particularly concerned about a serious gap in one fire service's ability to respond to a terror attack.... this must change..."*.

The variations in response configurations and staffing in the UK is currently so varied that it would be impossible to identify an objective national position for managing specific risks. A fire in a flat, for example, will receive a very different response depending on where in the country this incident takes place. It is understood that out of the 38 fire stations that Lincolnshire Fire and Rescue Service manage, only one is staffed by full-time firefighters on a 24-hour basis (Lincolnshire Fire and Rescue Service, 2018: p-3). This is compelling seeing as Lincolnshire in the second largest county in the UK (Doyle, 2011). Conversely, Buckinghamshire Fire and Rescue Service have established a bespoke response model to satisfy their response objectives for retained duty system firefighters[7].

[7] Retained Duty System firefighters are paid on-call firefighters. As opposed to full-time firefighters, retained firefighters have other principal occupations and provide availability for duty to their nearest fire station. They are alerted by pager and must respond to the fire station within a couple of minutes of

Response Model

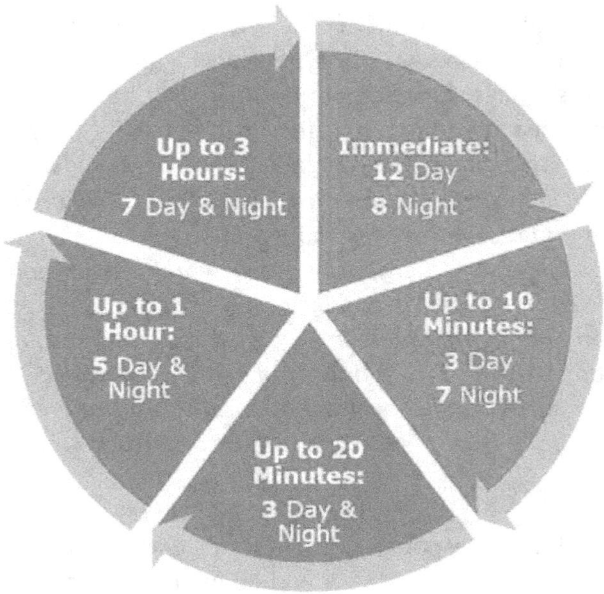

Key:
Green: Response Model for Business as Usual Demand
Amber: Response Model for Resilience

Figure 2 – Bucks Fire and Rescue Service Response Model for Retained Firefighters

(Buckinghamshire Fire and Rescue Service, 2020)

The availability configurations are denotated in the green and amber pies above. All retained duty system firefighters wanting

---

the pager alert. These firefighters are paid a retaining fee and remunerated on an hourly rate for their duties.

to choose to provide against the more flexible resilience cover options (Amber Pies) must also be in the position to provide a minimum immediate response availability of 38-hours per week (Business and Usual Demand – Green Pies) (Bucks Fire and Rescue, 2020). The Green Pies represent the standards required for immediate and 'up to 10 minutes' response cover. Therefore, if a retained firefighter was able to provide an 'up to 20 minutes' response availability for three days and nights, they would also be required to provide at least 38 hours of immediate response availability per week. Fire services such as West Midlands Fire Service and the London Fire Brigade, as part of their assessment around emergency cover, have established that in order to meet their operational objectives, they can only employ full-time firefighters.

This impact of austerity measures is not just restricted to the UK, though the evidence available presents some stark examples. Fire services from many other countries are also suffering the consequences of the financial crisis. In the United States for example, the documentary film *'Burn'* (2013) depicts the severe effects of austerity on the city of Detroit and consequently the Detroit Fire Department. In an article by former Fire Chief, Robert Rielage (2014), he reveals that as a response to the budget cuts in the city of Detroit, 'some of the answers from local government was to underfund capital improvement or equipment replacement funds. For the fire service this meant deferring or eliminating the replacement of worn fire and emergency medical apparatus, turnout gear, self-contained breathing apparatus or fire stations — even after it was proven that the investment would immediately save money and improve the department's long-term efficiency'. Additionally, it was identified that other local governments decided that firefighters — whether career, part-time or even volunteers — were too expensive, and subsequently reduced the number of personnel available for emergencies as they determined that staff was their largest expense[8]. In the documentary film, senior ranking officers of the Detroit Fire Department demonstrated with fact and figures, the minimum number of firefighters required to deal with the fire epidemic effectively and safely in the city Detroit. At the time of the documentary those numbers were far from adequate and being decimated even further.

Other reports from the United States have indicated serious concerns over budgets cuts made to fire service programmes by the Trump constituency over two consecutive fiscal years, 2019 and 2020. The International Association of Fire Chief's had highlighted the implications of these cuts that will likely affect fire departments to adequately respond to a variety of threats (La Sala, 2018)[9]. The United States Fire Administration (2019)[10] has expressed similar concerns, especially with regard to how budget cuts will affect their ability to accomplish their mission around data collection, public education, research and training, the purpose of which is to help reduce fire deaths and make communities and their residents safer[11].

In Germany, fire services are suffering from the lack of uptake of volunteers. The German fire service is unique in the sense that around 95% of its firefighters are volunteers. This is a long-standing tradition in German culture that boasts over 22,000 voluntary fire brigades in local townships, rural areas and on the outskirts of larger urban settings. Though, cities with more than 90,000 residents are obliged to have professional fire departments, these are often also supported by volunteers on the periphery (Isenson, 2019). However, it is argued that the economic crisis is taking its toll in Germany as well, as due to

---

[8] https://www.firerescue1.com/fire-chief/articles/5-ways-to-show-effects-of-fire-budget-cuts-QFWFMFIp0xGdFUeu/

[9] https://www.iafc.org/blogs/blog/iafc/2018/02/16/fire-service-programs-cut-by-president-trump-s-fy-2019-budget

[10] https://www.everycrsreport.com/reports/RS20071.html

[11] https://www.usfa.fema.gov/about/index.html

financial constraints people are disincentivised, or less likely, to offer their services as volunteers. Whereas nearly 1,070,000 people were members of voluntary fire brigades in 2000, by 2016 that figure had fallen to around 995,000, according to the German Firefighters Association (DFV). As argued by Carten-Michael Pix of the DFV *"the voluntary fire department is a mirror of society. German society is also getting smaller and is struggling with the same problems as we are. That is one of many reasons why the number of firefighters is falling"*.

Similarly, in the Netherlands, 80% of firefighters are part-time. In his post graduate thesis, 'Volunteer Firefighters and their intention to Remain Active', Tamo Vogel (2014: p – 18) discusses a number of reasons why people join and leave the fire service. Amongst those was the growing bureaucracy and regulations that limit individual freedom and therewith the pleasure in their work. Haverkamp (2005) also concluded that an increase in work and time burden is an important reason for volunteers to stop. The support of employers, partners and authorities are indispensable for actual availability of volunteers. Most importantly, however, it was identified that because of the economic crisis, municipalities have been confronted with budget cuts from the central government and a loss in revenue, fees and reserves. In most regions of the Netherlands this has also affected the budget for the fire service. In certain circumstances this has already resulted in the reductions of budgets, disposal of fire trucks and in some cases even the complete closure of fire stations. Some regions or municipalities also cut back the compensation of their volunteers (Vogl, 2014: p – 19).

Similar crises with marked reductions in volunteers are occurring across the world. In 2009, 71% of North American firefighters were volunteers (NFPA, 2010). By 2019 this number had reduced to 65% and it is reported that in 2019, volunteer firefighters were at an all-time low across the country, with the average age of firefighters increasing significantly whilst the call volume to emergency incidents has tripled in the last 30 years (National Volunteer Fire Council, 2019: p - 2).

In Greece, in 2008, one of the first budgets to be cut in response to austerity measures was the annual funding for civil protection volunteers. Until that time and from around 2002, civil protection volunteer organisations were supported by the ministry of interior with small budgets to assist in the purchase and maintenance of operational equipment. As the criteria for these funds and how they were used was questionable, and with a number of concerns around the misappropriation of such funds, it made sense for these to be axed as a matter of priority (Pantelides and Tziritis, 2009).

Unfortunately, this also had an impact on many active volunteer organisations that have had to resort to raising funds through other mean and even paying out of their own pockets on regular occasions to support their operational capabilities. This has had a significant impact on the development of many such organisations who are unable to maintain their existing equipment (or purchase from new) which is often old, outdated and in poor condition. On the other hand, the budget for the entire Hellenic Fire Service in 2017 was equivalent to that of the London Fire Brigade alone (Pan-Hellenic Contract Firefighters Union, 2017). Similar stories of budget cuts and ageing workforce appear prevalent across the board. The multitude of online articles paint a clear picture of the problems that fire services around the world are facing as a result of austerity.

The financial crisis and the measures required for cost-saving present a multitude of issues that impact directly on effective service delivery of fire and rescue services internationally. Lower staffing numbers and response configurations, slower response times, acceptance of increased risks, the proliferation of alternative employment structures with volunteers, part-time firefighters, flexible-duty system firefighters and the replacement of traditional 24-hour fire stations; the reduction in the number of fire stations, fire appliances; the retention of older equipment with less investment on maintenance and upgrading; staff redundancies; the merging of services and roles to capitalise on cost-efficiency; the undertaking of additional work such as responding to first-aid incidents along with the ambulance service. These are just some of the implications of the financial crisis and consequential budget cuts. It has already been discussed earlier that big-data analysis can provide essential insights into almost any area of interest that can be measured. Of course, adapting to change is part of fire service development and, in many cases, this can be qualified by official metrics. There is value in change, but care must be taken so that overreliance on the numbers alone does not serve to govern judgement and decision-making processes to satisfy organisational objectives even when there is evidence to demonstrate an increased risk to firefighters and the public.

It is suggested that it is in the nature of fire services as organisations to overcome difficulties and challenges that arise in the field. It is thus assumed that they are more resilient to circumstances where they are required to do more with less. However, it is argued that this is neither safe nor risk proportionate. Dr Steven Knight, who is the fire service practice lead and partner at the public safety consulting firm, Fitch & Associates, asserts that *"the first step in a discussion on fiscal constraints is to understand the modern public fire safety department's budget. As public safety is a labour-intensive service model, typically, more than 90 percent of the budget is accounted for by personnel costs. An additional 5-7 percent is usually assigned to fixed costs for items such as apparatus parts and repairs, fuel, hose, equipment, station supplies and station maintenance. Ultimately, there is typically less than 5 percent discretionary spending capacity that frequently covers unexpected expenses. Therefore, incremental cuts of only a percent or two can have a significant impact on a department's ability to maintain services, because discretionary spending is so limited. In other words, there may not be sufficient cushion to insulate adjustments without impacting personnel and/or service delivery".*

The Chief Fire Officer's Association UK report in 2016 strongly asserted similar points. Findings from their survey around 'doing more or the same with less' in 2012 identified that:

- Almost all FRSs have already reduced the number of frontline firefighters.
- 90% of Chief Fire Officers (CFOs) believe that further cuts in front line posts are inevitable.
- 97% of FRSs have cut back-office staff and three quarters have implemented redundancy programmes amongst corporate or back-office staff.
- 90% of CFOs do not believe that sufficient savings can be made through changing their back-office support functions, better procurement or sharing senior teams. The savings required to set balanced budgets are just too big.
- Almost half of FRSs expect to reduce the number of fire engines and a quarter are planning to close fire stations.
- 95% of CFOs expect to see an impact on national resilience (civil protection).

Change does not come without consequence, and it is inevitable that certain changes will promote certain limitations. Indeed, it would perhaps not be financially viable to finance a 24-hour fire station with full-time firefighters in an area that has low incident activity. However, problems arise where there is no objective benchmark to drive such decisions i.e., at what number of incidents would a fire station, other than a full-time fire station, be warranted? When is it safe to send out a fire appliance with a 3-person crew as opposed to 4 or 5? Some fire services have specialised response vehicles for 3-person crews that have been established to respond to lower-risk incidents (rubbish fires, lift entrapments, false alarms etc.). There is value in utilising such response configurations where this can be justified, risk assessed appropriately and with the appropriate caveats in place, but these configurations are not implemented with such forethought across the board. DeShon, (2016)[12] argues that *"in the case of personnel, it has been proven time and again that we* cannot *do more with less. In fact, when understaffed, we are not only inefficient, we are also more likely to be injured"*. The research around such matters is prominent and will be discussed later. DeShon further asserts that *"in busy departments where tight budgets require firefighters to do more with less on a daily basis, these risks are often multiplied by an extreme call load and chronically fatigued firefighters, posing a real risk to the communities they serve"*.

The underlying problem is that the variables that govern these decisions currently, in the UK, and other countries, are too ambiguous, meaning that quality of service and acceptance of risk are increasingly subjective. In the UK, the move from national standards to localised risk assessments to govern service delivery objectives allowed for this level of diversity across the risk management spectrum where different fire services are doing different things to manage the same types of risk. However, it is also suggested that older national standards could not ensure adequate levels of service either and the balance appears to lie somewhere in the middle.

Further consideration must be given to the fact that, if fire services, such as those in Northern Europe and other countries that have been industry leaders, are susceptible to the impacts of austerity despite their robust systems, policies and procedures, then how are other fire services with far less robust

[12] https://firerescuemagazine.firefighternation.com/2016/03/01/can-we-do-more-with-less/#gref

organisational structures going to provide quality assurance in their service delivery? This is perhaps one of the key downfalls arising from the absence of benchmarks despite the plethora of research and evidence that indicates a series of best practice methods for achieving organisational and operational goals.

Lessons from the International Fire Instructors Workshop.

In June 2019, the author was invited to take part in the International Fire Instructors Workshop (IFIW) at the Institute for Safety (IFV) in the Netherlands. This was followed by the two-day International Fire Instructors Conference (IFIC) that was an open event. The IFIW, that took place on the first three days, comprised a closed group of fire service specialists ranging from fire safety engineers, fire behaviour instructors, fire scientists, academics and so on. The contingent does not distinguish between rank and status and, therefore, hosts a range of individuals that are both professionals as well as volunteers.

The year's topic of discussion was to establish a series of 'non-negotiables' on the fire-ground[13] for structural fire events. A couple of months prior, those invited to attend the IFIW were asked to submit their own non-negotiable via email to the organisers. This created a pool of over 20 - 30 items that were selected to be discussed at the workshop. The objective of the IFIW was to conclude on eight universal non-negotiable factors that could be worked on and eventually promoted as a benchmark standard. Over the three days of the IFIW, eight factors were selected from a pool of over twenty. There were some interesting themes that arose from the discussions.

[13] Fireground refers to the location of a fire incident where firefighters are undertaking operational duties.

Firstly, it was established that the capability around the practical application of non-negotiables was contextual. Many fire services from northern Europe and the United States were starting from a firm standing point, where a plethora of procedures were already in place to support the further promotion of these concepts. The context for Greece and, to a lesser extent Spain, which was also represented, was quite different. An example of agreed 'non negotiables' was that every crew inside a burning structure must have a thermal imaging camera and knowledge of how to use it correctly. Another was that firefighters should always undertake action to cool the smoke and another opined that firefighters should identify the flow paths and close all openings prior to fire control. Such non-negotiable factors would commonly be embedded in standard operation procedure or a guidance document that is supported by adequate training.

In Greece, such processes are not even covered in basic firefighter training let alone documented in a procedure or guideline[14]. Thermal imaging cameras, for example, were only purchased circa 2016, are limited in number and are not used extensively. As such, the application of non-negotiables is fundamentally dependent on whether a fire service has the appropriate structure and ability to adopt them. It was made clear from the Greek representatives, and to an extent those from Spain, that most of the non-negotiables were not realistic from the point that there were no foundations to support them in the first place in their respective fire services.

However, if representatives from more than a dozen fire services (16 different countries were represented at the IFIW)

[14] Although there is a training programme that intends to introduce these concepts that is primarily geared around fire behaviour training in the bult environment, the programme is rolled out to existing firefighters, many of which do not have the foundational knowledge or willingness to adopt these procedures in practice.

are able to broadly agree on a series of operational rules and processes in the form of non-negotiables, then this would indicate that the logic of a benchmark standard is potentially achievable. Though this was inferred in discussions with many fire service representatives agreeing on operational processes and procedures, a benchmark was not explicitly identified. It was also apparent that many fire services have different methods of achieving the same goal, none of which are necessarily better or worse than the other. Perhaps a commitment to certain methods of practice at organisational level makes the process of establishing a universal benchmark difficult, as this require change which is not overly popular in the fire service world (Gasaway, 2009)[15]. It could be argued that a benchmark should in fact be set against the fundamental goals that are to be achieved by a fire service and not necessarily against the methods of achieving those goals, providing they are both sound and safe for firefighters, members of the public and supported by scientific evidence and practical experience.

There were a number of other issues discussed at the IFIW that identified commonalities across almost all fire services. One of the key issues was regarding the practical application of firefighter training. Since the 1980s, there have been developments in fire extinguishing techniques. Though sometimes fighting fires inside structures is as simple as 'putting the wet stuff on the hot stuff', it has been recognised that it is essential for firefighters to understand the basic physics of fire behaviour and water application, the effects of ventilation and the management of flow-paths etc. in order to be more effective in dealing with compartment fires.

John McDonough from New South Wales Fire and Rescue Service has been a world-renowned fire behaviour training instructor for over 20 years. At the ensuing International Fire

---

[15] https://www.samatters.com/explaining-resistance-change/

Instructors Conference, McDonough spoke about his recent transition back to operational shift-work and how surprised he was to find that firefighters at fire incidents will, in fact, simply revert to habitual fire extinguishing techniques, despite their training.

Firefighters learn by application, repetition and experience (Sandlebeen, 2002; Denef et al., 2008) and if what they do is working, then they will often continue to use that method as standard practice. Amongst other things, fire behaviour training teaches firefighters how to prevent rapid fire development phenomena in order to remain safe from an extreme event such as a flashover, backdraught or fire gas ignition. This utilises gas cooling techniques and ventilation control to achieve these outcomes.

However, firefighter fatality and injury statistics would demonstrate that out of all the interventions that the fire service attends, the cause of injury or death to firefighters from such phenomena is in fact very low. Indicatively, the NFPA reported that in 2017 they experienced only 17 deaths at the scene of fires and only nine of those were at structural fires[16]. Similarly, in the UK, it is reported that between 1986 and 2018, 69 firefighters died whilst on duty and 28 of those were at fire incidents[17].

With this in mind, it becomes increasingly difficult to convince firefighters that what they have been doing for decades that has worked previously at every other incident alike, might not work under certain circumstances. However, in firefighter fatality case studies from the UK, such as Blaina in 1996 (Fire Brigades Union, 1996), Bethnal Green in 2004 (FBU, 2004), Shirley Towers

[16] https://www.nfpa.org/News-and-Research/Publications-and-media/NFPA-Journal/2018/July-August-2018/Features/Firefighter-Fatalities

[17] https://data.gov.uk/dataset/f7ccba2d-f679-46e1-8d6d-de7bee690d40/fire-statistics-fatalities-and-casualties

in 2010 (Hampshire Fire and Rescue Service, 2013), it has been determined that firefighters were not adequately prepared to deal with conditions arising from rapid-fire development and perished as a result. These events though are too few and far between to prompt firefighters to consider that a different approach might have resulted in a more favourable outcomes. McDonough (2019) specifically argued that *"gas cooling trains firefighters for 2% of fire incidents"*[18]. The figure of 2% is indicative and will realistically vary from country to country. The point McDonough is making is that although gas cooling techniques should be applied as standard practice in hot-smoke conditions it is often not. Most of the time, the conditions for gas cooling are indistinguishable across the continuum and firefighters generally appear to get away with not gas cooling until those very rare occasions where it is essential to gas cool and they have not done so. It cannot be denied that there is merit in traditional methods of firefighting, after all, firefighters are attending fires and putting them out, and evidently without event in the vast majority of cases.

An interesting discussion occurred around the non-negotiable 'stay-low'. In principle, this is good practice, as the atmosphere

---

[18] Gas cooling is a firefighting technique used to cool the hot fire gases in the smoke layer of a burning structure. It is based on the concept that smoke is fuel and is therefore flammable. If the smoke ignites inside a compartment involved in fire, the temperature within the smoke layer will be greater than the ignition temperature of all other fuel sources within the room (mainly furniture). Once flaming combustion occurs in the smoke layer, all of the materials in the room will ignite, almost simultaneously, from the radiant heat generated in the smoke layer, generating huge amounts of heat. This is called flashover from where a fire transitions from its growth stage into a fully-developed fire where the materials inside the room can release their maximum heat potential. These condition are untenable for firefighters. A precursor to flashover is something called rollover. This is indicated by strips of flame propagating through the smoke layer igniting the hot gases. To avoid this sequence of events, firefighters apply short bursts of water into the smoke layer using small water droplets that evaporate displacing the available oxygen, reducing heat and the flammability of the gases in the smoke.

is cooler, and visibility is often better. However. It was argued by some that 'stay low' should apply under certain circumstances – not always. For example, in buildings with high ceilings (such as warehouses and factories) where the smoke layer is usually at high level, above head height, (due to the tall ceilings), staying low would not satisfy any purpose and serve to delay advancement to the seat of the fire. In calls regarding food that has been left on the stove, it would be of no value to enter into this incident staying low. Staying low would apply where there is limited visibility and hot fire gases within the smoke layer that are forcing firefighters to stay low in more viable conditions.

Some firefighters agreed that staying low is appropriate once hot conditions are confronted. A counterargument to this was that if firefighters wait until it is too hot before they go low, then it is already too late as their personal protective equipment will have been saturated by the heat, enough so to penetrate the thermal barriers of their protective uniforms and potentially cause discomfort to the extent that it would require firefighters to retreat or even cause them burns. As both arguments are perceived as equally founded, a different perspective would be to train firefighters to acquire the expertise to determine when staying low is required and when not. Unfortunately, there is plenty of evidence from fire firefighter Line of Duty Deaths to demonstrate that this can also be problematic as personal judgement is not always sound and certainly not always objective (Reason, 1997; Moore-Merrell et al., 2006; Siarnicki, 2006).

Another long-standing debate between fire services (especially between Europe and the United States) is with regard to gas cooling versus straight (or solid) stream water application, combination nozzles versus smooth bore nozzles and so on. The one side argues that gas cooling with the use of fine water droplets in an open spray applied in short bursts into the hot gases is the correct approach for creating safer conditions when advancing towards the fire. These techniques serve to cool hot gases by evaporating the fine water particles into steam, absorbing large amounts of heat in the process, inerting the flammable gases, displacing the available oxygen and contracting the hot fire gases within the compartment (Svensson and Van de Veire, 2018). This method favours the use of a combination nozzle where both a straight stream and a fog pattern can be created and where adjustments can be made to the capacity of the water output in litres per minute. The other side is in favour of smooth bore nozzles that provide a solid stream allowing for greater reach and higher flowrates of water directly applied onto the fire and hot surfaces. An Underwriter's Laboratories study 'Impact of Fire Attack Utilizing Interior and Exterior Streams on Firefighter Safety and Occupant Survival: Water Mapping" argued that fire departments base nozzle choices and hand-line set-ups (hose line size, desired nozzle flow, required pressure) on existing knowledge, traditions, and beliefs (Weinschenk et al. 2017: p – 4). It is apparent that these beliefs are not often supported with scientific evidence.

Chief Ed Hartin, in 2013[19], provides an insightful response to Jason Sowder's article 'Nozzle Selection: Are we Defeating the Enemy' (2013)[20] in the online magazine 'Fire Engineering'. Ed Hartin is currently Chief of Central Whidbey Island Fire and Rescue in the state of Washington, USA. He is an internationally renowned fire service expert with numerous publications around fire service tactics and training. At the time of writing his article, Jason Sowders was company commander and Emergency Medical Technician (EMT) with Bowling Green Fire Department in Kentucky, USA. He also was serving as a training officer for his fire department. In his article, Sowders takes a critical view on the use of combination nozzles, gas cooling and fog patterns with a specific bias towards the use of smooth bore and straight stream

---

[19] http://cfbt-us.com/wordpress/?tag=indirect-attack

[20] https://www.fireengineering.com/articles/2013/06/nozzle-selection--are-defeating-the-enemy-.html#gref

attack. He specifically mentions in his article that; *"most fire departments throughout the country are aware of the harmful effects of fog application and are teaching their recruits to use straight stream water application for interior structural firefighting"*. In response to this article, Hartin (2013) provides a series of counterarguments that are based on scientific evidence and research undertaken into specific issues raised by Sowders. For example, Sowders states that *"We have been fooled for many years believing that a curtain of water between you and the fire is protection. What is occurring is that you are pushing heat, fire, smoke, and other products of combustion out in front of you"*. Hartin responds as follows *"As demonstrated in tests conducted by Underwriters Laboratories (UL) on Horizontal (Kerber, 2011) and Vertical Ventilation as well as additional tests conducted by the National Institute of Standards and Technology (NIST) and the Fire Department of the City of New York (FDNY) (Healey, Madrzykowski, Kerber, & Ceriello, 2013), water does not push fire"*. Throughout his response, Hartin provides sound arguments supported by scientific evidence effectively refuting Sowder's statements. He is promoting the position that, these cases are neither black nor white and that both a fog pattern for gas cooling and straight (solid) stream for direct attack, nozzle selection and flowrate are equally useful depending on the circumstances. Hartin closes his article with the following conclusions:

> *"...our strategies, tactics, and doctrine must be evidence based, must have a sound theoretical foundation and be supported by both scientific research and practical experience. Unfortunately, our profession continues to struggle to integrate these elements and is saddled with conclusions based on experience without understanding. Theory and scientific research do not trump experience, neither does experience trump scientific knowledge. Both are essential. The issues of flow rate and stream selection are not one sided, there*

is evidence for the effectiveness of both water fog and solid stream application for control of fires in today's fire environment. It is easy to examine the evidence and choose the facts that support our preconceived ideas (regardless of your perspective). It is much more difficult to objectively evaluate the evidence and determine what conclusions are actually supported. We must continue to ask why and question our assumptions".

(Hartin, 2013)

Another article by John McDonough who took part in the UL research (mentioned above) as part of the technical panel identified in his article, 'the Science of Water Extinguishment' (2018)[21] some issues that would affect the study in having global applicability. He mentioned that the flowrates selected for this study were as much as six times higher than the flowrates used in Europe and Australia whereas the nozzle types selected (smooth bore) are practically non-existent or rarely ever used outside North America. Conversely, he identified that other countries around the world still manage to extinguish fires effectively with far less water placing greater emphasis on technique rather than capacity. Though he agrees that larger amounts of water will serve to extinguish most common fires, McDonough identifies that high flowrates are not required at every fire and that there are consequences, such as water damage, that fire services should strive to avoid where possible. He argues that skill through effective water application tactics are fundamental requisites for a professional and effective fire service.

It further became apparent at the IFIW that fire services from around the world have their own intrinsic problems that must be understood within a specific context. These intrinsic issues will

[21] https://www.firehouse.com/operations-training/hoselines-water-appliances/article/12395735/the-science-of-water-extinguishment-international-innovations-john-mcdonough

unequivocally impact service delivery to some extent. Some preconceived ideas around fire service structure and delivery and the inadequacy thereof that prompted this study appear to not solely be restricted to countries suffering the brunt of the financial crisis, such as those in Southern Europe. Though there was a broad baseline insofar as the provision of policies and procedures around incident command, health, safety and well-being, maintenance of vehicles and equipment, standard operational guidelines for different types of incidents, fire, accident investigation etc. the methods of achieving these varied between fire services, quite significantly in some cases.

UK fire services appeared to have the most robust accountability process for firefighters entering burning structures, whilst some had a variation of this model and others did not have a standardised process for accountability at all. The crew sizes varied significantly between fire services, though most of them understood the necessity for adequate staffing numbers. Afterall, there is research to demonstrate its value which is mentioned later in This book. Another common theme was that when discussing various views on operational matters, especially those around the selected non-negotiables, it was established that fire incidents are too dynamic in nature for there to be fixed and definitive responses. A universal phrase used more often than not in response to most issues discussed at the IFIW, and more broadly in fire service language, was 'it depends'.

Indeed, the possible scenarios that a firefighter is likely to face in any given situation can be so varied that it would not be sound to create a set rules that are too rigid to allow for adaptation to changing circumstances. As such, as a general consensus it was believed that firefighters should be provided with a set of tools in the form of guidelines and be trained to make effective decisions depending on the situation and present conditions. Some believed that these guidelines need to apply a level of rigidity so that firefighters are not completely free to make decisions that could ultimately put them and their crews at risk.

Although it is established that there is a significant level of diversity between fire services from around the world, many in the more developed countries share commonalties in the ethos and priorities for achieving service excellence, though the methodology on how these are achieved between each service can vary and, in some cases, significantly. It can be argued that where variation does occur this will not raise question to the reliability providing that the methods in achieving the end goal are supported by sound scientific evidence in conjunction with lessons obtained from practical field experience. Problems arise when the methods towards achieving a specific goal are not supported with strong empirical evidence and where emphasis is placed on experience alone or vice versa. This will undoubtedly hinder the development of a fire service and differentiates between those leading the way forward in the fire service world. As suggested, there are some common denominators that are representative of modern fire services. These have to do with ensuring adequate staffing numbers, the standardization of training practices, the importance of using standard operating guidelines, health, safety and accountability for firefighters and so on. The following sections will address these fundamental principles in turn which should be viewed as forming the framework for the development of universal baseline standards.

In the next section, some key aspects that are pertinent to assuring fire service effectiveness will be discussed. These are matters such as the necessity for adequate resourcing, the importance of standard operating procedures and guidelines, the role of the incident commander, health and safety and so on. This, of course, is not an exhaustive list, but merely a sample of indicators to demonstrate the criticality behind establishing a structured approach towards ensuring service delivery. These elements should apply to any fire service regardless of geographic location as the commonalities of these elements are what make fire services an international profession - within context. The above argument is further supported by two separate studies conducted on behalf of the International Association of Firefighters. The first, identified that some of the main contributing factors leading to fire service line of duty deaths (LODDs) in the United States between the years 2000 – 2005 were inconsistencies in crew size, heath, wellness, fitness and medical status of firefighters, as well as non-conformance to, or lack of, standard operating guidelines (Moore – Merrell et al, 2006: p 33). The second study by Moore – Merrell et al (2008: p 28), into line of duty injuries of firefighters at Metropolitan Fire Services in the United States between 2005 – 2006 found that the main contributing factors leading to injury were decision making, lack of communication, standard operating guidelines or procedure breaches, protocol breaches, human error, lack of situational awareness. The importance of these salient points will be discussed in the next section.

Chapter Summary

This literature review has explored the multidimensional challenges and opportunities shaping the development of fire and rescue services, with particular attention to organisational learning, data utilisation, and cultural dynamics. While technical competence remains essential, the chapter has demonstrated that the performance and reform potential of fire services are dependent upon their institutional capacity for reflection, adaptation, and learning. The comparison between North and South European contexts highlights stark disparities in organisational maturity, with the former generally characterised by systematised governance, evidence-informed practice, and integrated data systems, and the latter often constrained by political instability, cultural inertia, and limited resourcing.

A recurring theme across the literature is the tension between formal organisational frameworks and informal, often habitual, operational practices. This friction manifests in procedural deviations, resistance to change, and risk normalisation, which persist despite the implementation of robust policies and procedures. Furthermore, while advances in data analytics offer immense potential and insight towards establishing performance targets and an organisation's effectiveness in meeting those, their impact can be uneven and influenced by organisational culture, leadership receptivity, and the political instrumentalization of data.

Crucially, these structural and cultural asymmetries present crucial obstacles to the development of universal baseline standards across European fire services. Without shared institutional capacity, mutual accountability frameworks, and culturally embedded learning practices, efforts to harmonise quality assurance systems appear intangible. Standardisation cannot rely solely on outcome based metrics and must be underpinned by an alignment of values, governance models, and institutional competencies.

In sum, the literature review emphasizes the necessity of embedding a learning ethos within fire service organisations—one that transcends compliance and performance rhetoric and cultivates genuine organisational intelligence. Achieving this, however, will require not only technical and procedural reforms but also a sustained interrogation of the socio-cultural and political logics that underpin fire service practice, particularly in underperforming or structurally disadvantaged systems. This sets the stage for a more critical and context-sensitive inquiry into the implementation and efficacy of quality assurance systems in European fire services.

Chapter 3 – Defining a Thematic Framework for Evaluating Fire Service Effectiveness

The research framework forming the quantitative part of this study was structured against several key themes. These are in relation to the resourcing of fire and rescue services, staffing levels and crew consistency, the adequacy of the provision and maintenance of personal protective equipment, the existence of, and conformance with Standard Operating Procedures, Incident Command structures, Firefighter Health and Safety and the importance of Radio Communications. Each one of these aspects is critical to effective, efficient and safe fire service operations. These themes form the foundational basis of benchmark standards to which fire services in this research have been assessed against. A minimum standard (or set of basic principles) has been established from each of these themes (e.g. all firefighters should wear their personal protective equipment at emergency incidents, all fire service operations should be supported by operational procedures or guidelines etc.). These themes will be discussed, in turn, to provide better context to deviations drawn from the responses of fire services included in this research and further underline the implications arising from a failure to meet the minimum standards purported in this study.

Resourcing of Fire and Rescue Services

Fire service resources are generally more heavily focused in the cities due to the large numbers of people, businesses, dwellings and the extent of processes and activities taking place there that present an inherently higher risk of fire service related incident occurring (Chung et al., 2024). On a risk rating scale, it is mainly the likelihood of an incident occurring that is in fact greater when compared to outer-urban or rural areas, although based on circumstance, there is also potential for the severity in terms of consequence to be extensive if not dealt with effectively. As such, it appears that the management of risk and the allocation of resources are generally structured around demographic factors (Ezlamzadeh et al., 2022).

Indeed, where more people, buildings and processes are present, the likelihood of an incident requiring an emergency services response will be higher than in less densely populated areas. In an attempt to reduce this risk, many metropolitan fire services will implement response time targets for the arrival of fire engines at an incident following the initial call for assistance. Response times outside metropolitan areas are generally greater than those in inner city areas. This comes as a result of more sparsely located fire stations that in many countries tend to utilise on-call firefighters or volunteers. It is not uncommon for fire services in many countries to travel tens of miles to reach the most remote areas of their response ground as identified in the data analysis section of this research.

However, the severity, or consequences of risk to a secluded ski resort with 200 guests which will warrant a slower response time and will further draw comparatively fewer resources than in the city, would have the potential to be greater. Nonetheless, a hotel with 200 occupants in the city will, most probably, receive a quicker, faster and better response than the secluded ski resort, despite having the same risk profile. Objectively, the overall service provided to the hotel in the city will be better than the one provided to the ski resort and this is commonly compensated for by the provision of enhanced fire protection measures such as sprinklers (Gernay et al., 2023).

It is here that more economically efficient models could be considered, such as the utilisation of volunteers, retained on-call staff and mixed or day crewed fire stations. There is also value in developing more robust fire safety procedures that require enhanced life safety features to be incorporated into buildings at the design stage or even retrospectively. Fire safety measures, such as sprinklers, automatic fire detection and other active and passive fire safety measures should be required through legislation in order to assert the level of importance around their implementation and need to be enforced accordingly. The ongoing management of fire safety must then be maintained through appropriate training and, collectively, these features will serve to minimise the likelihood of a fire occurring but also the consequences should a fire occur.

As argued by Paul Grimwood in a recent interview (2018), *'where fire service intervention cannot be guaranteed, the shortfall must be supplemented by robust fire safety measures and supported by an adequate system of management'*. Thus, in a balance of cost-efficiency, enhanced fire protection measures would compensate for slower fire service response times and lesser resources. Such an example can be seen at St Michael's Mount in Cornwall. The castle that sits on top of St Michael's Mount is at increased risk of fire damage as a result of the access issues regulated around high and low tide. During high tide the castle is not accessible by road and fire engines cannot reach the castle boundaries even during low tide. As the castle is open to the public, the risk of arson cannot be excluded. As such, there are specific local management arrangements to ensure that a fire can be kept in check until the fire service arrives. These measures have to do with the implementation of robust safety features provided to the castle in the form of a large water tank and high volume pump that can fill the water mains system located strategically around the castle, the installation of a comprehensive fire detection and alarm system, the adequate training of staff, the enhanced fire safety features (fire doors and compartmentation) and a hybrid amphibian vehicle that can be used on land and water to transport firefighters from the shore to the castle dock (The Packet, 2015).

Many fire departments around the world provide emergency cover through a variation of models where full-time paid firefighters providing 24-hour emergency response are pre-dominantly located in inner city areas whilst other forms of cover are provided outside of these. In Germany, for example, in 2021 there are 35,875 full-time firefighters providing emergency response cover out of 111 city fire stations and 1,014,155 volunteer firefighters providing cover to their local communities out of 23,977 fire stations (Deutscher Feuerwehr Vebrand, 2021). In England, in 2023 there was a total of 30,723 firefighters providing emergency response cover. Out of these, 22,626 were full-time staff and 10,219 were retained duty system firefighters[22]. It is interesting to see the distribution of staff by metropolitan or non-metropolitan area, simply to reaffirm the previous point that full-time fire cover is the primary model adopted in inner-city areas. This is demonstrated below:

FRS	Wholetime	Retained Duty System	Total Firefighters
England	22,626	8,097	30,723
Metropolitan fire and rescue authorities	9,936	283	10,219
Non-Metropolitan fire and rescue authorities	12,689	7,814	20,503

Table 1: Distribution of wholetime and retained firefighters in the UK (2023)

(Home Office Fire Statistics Data Tables, 2023)

This point is further supported by Haynes and Stein (2017) in a report for the National Fire Protection Association titled 'U.S. Fire Department Profile – 2015' that identified that most of the career firefighters (71%) in the United States, worked in communities that protected 25,000 or more people whilst most of the volunteer firefighters (95%) were in departments that

[22] https://www.gov.uk/government/statistical-data-sets/fire-statistics-data-tables#workforce-and-workforce-diversity

protected fewer than 25,000 people. In Germany, any city with more than 90,000 residents must have a professional fire station (Isenson, 2019).

Similar examples can be seen in the Table 2:

Country	Full-time	FT Depts or Stations	Volunteers	V- Depts or Stations
New South Wales Australia	6,822[23]	335	72,855[24]	1,993
United States of America[25]	364,300	5,300	676,900	24,150*
Portugal[26]	2,000	17	40,000	467

**There are over 50,000 fire stations in the United States of which the majority are volunteer fire stations.*

Table 2: Distribution of wholetime and volunteer staff by country.

In summary, it appears that 24-hour full-time fire stations are more widely represented in inner urban areas. This is due to the frequency and risk profile of incidents and the proficiency that firefighters are required to have in dealing with such incidents on daily basis. Outside the cities there is a variation of different models of fire cover that include part-time firefighters, volunteers, day-crew stations and so on. These models are more financially viable and in northern Europe and the UK, the absence of operational experience is replaced with frequent and realistic training. This is not necessarily the case for other fire services in southern Europe as has been identified in the data analysis section of this research.

The impact of the financial crisis has prompted numerous fire services to review their provision of emergency cover and find innovative ways of cost-saving. Such practices have considered down-sizing crews, replacing less busy full-time stations in cities with on-call firefighters, creating flexible shift patterns as well as integrating other resourceful adjustments. Though these methods can assist with cost saving, there is contention on whether the quality-of-service delivery is being maintained or in fact declining. Other fire services appear to start from a different baseline meaning that three-person crews (or less) or greater response times without a contingency are the norm at organisational level and, therefore, there is very little room to manoeuvre in such scenarios. Crew size and operational experience play a key role in service delivery. At least in the case of crew size, there is plenty of research to demonstrate that staff numbers are vitally important to fire service operations.

First Response and Crew Consistency

The adequacy of crew consistency in the first response to any emergency incident plays a critical role, both in the extent and quality of the service that can be delivered, as well as ensuring the safety of the crew itself. It has already been discussed earlier that there is a direct association between the structure of fire service operations and the extent to which a service is able to be provided i.e., the more structured the approach, in terms of strategic robustness and firefighter competency, the safer crews are able to operate. This, in turn, leads to greater reliability in the service that is being delivered, and in the desired outcomes of a particular incident.

Paul Grimwood was a retired fire service veteran who served most of his 35-year career in the some of the busiest fire stations of the London Fire Brigade. He is widely recognised as one of the most influential fire service experts in the world with significant contributions and authorship regarding firefighter safety and the modernisation of tactical firefighting operations, with work dating back to the 1980s. It is accepted that his important work has contributed to the formulation of many of the operational procedures used contemporarily in the London Fire Brigade and beyond (Dudeney, 2018; Lambert, 2019).

In his book 'Euro-firefighter' (2008: pp 139-140), Grimwood discusses the importance of establishing crew capabilities at structural fire incidents. He argues that the different elements that must be addressed during structural firefighting operations will govern the number of firefighters that will be required in the primary (and secondary) response to adequately cope with the demands of such events.

Extensive research in the United States (Cushman, 1981; McManis Associates and John T. O'Hagan and Associates, 1984; Morrison, 1990; Roberts 1993; Verone, 1994) has measured staffing levels on fire engines, examining the effectiveness of crews of three, four and five firefighters attending structural fires, and other incidents. The results of these studies eventually led to the development of NFPA 1710 *'Standard for the Organization and Deployment of Fire Suppression Operations, Emergency Medical Operations, and Special Operations to the Public by Career Fire Departments'* in 2001 that recognised that adequate staffing is the most critical factor in effective and efficient fire service operations. Later research by Fire2000.com carried out in 2005 by an international team of operational fire officers and compartment fire behaviour training specialists with vast experience in the field derived the Critical Task Performance Index. The Index served as a competence grading for first arriving crews at structural fires and endorsed the research findings from the United States by pointing out that fire-ground performance and firefighter safety was severely compromised in situations of limited resource (Grimwood, 2008: pp 140).

This logic equally applies to all other incidents that are dealt with by the fire service, whether a fire, a road traffic collision or, for example, a water rescue. There are critical tasks that must be undertaken in the primary attendance at any incident that would command an adequate response in order to manage the requirements safely, efficiently and effectively. Thereafter, depending on the nature and extent or severity of an incident and on how it progresses, an exponential increase in resources might be required to satisfy operational demands.

Taking staffing numbers, for example, a fire contained to one room of a detached single storey dwelling might require, as a minimum response, two firefighters who would be involved in firefighting operations; an incident commander to oversee the requirements and progress of the incident in conjunction with the effectiveness of the chosen operational strategy; a driver or pump-operator to secure that firefighters had an inexhaustible supply of water (where possible) and at the correct pressure to deal with the fire. Some services might include a breathing apparatus entry control officer to ensure the accountability and safety of the firefighting crew working inside a building. Another firefighter might also undertake miscellaneous tasks between the driver - pump operator and the firefighting crews. However, if people were reported as trapped inside the structure and perhaps the fire was in a block of flats with a possibility of fire spread to other premises, then resources would have to be increased significantly to reflect the demands of that incident. More firefighters, mostly in teams of two, would have to be deployed to undertake[23]:

- Search and rescue operations.
- Firefighting activities (to prevent the spread of fire etc.).
- Back-up crews to protect firefighting crews.
- Pumping operations.
- Tactical ventilation.
- Rapid Intervention Teams to perform firefighter rescues if required (in some countries).
- Crew relief.
- Miscellaneous duties (carrying equipment, laying hoses etc.).

[23] It is acknowledged that not all fire services will necessarily staff these posts or might have less staff undertaking a combination of roles. However, this example is simply to demonstrate further resource requirements that might need to be addressed at a larger incident.

- Sector commanders to co-ordinate different sectors of the incident.
- Incident Command.

A variety of different type of specialist vehicles might be required such as aerial platforms, turntable ladders, incident support units, incident command units, hose-laying vehicles, canteens and so on. Similarly, even a single vehicle road traffic collision with one person trapped would require a team approach to cover the 'six key elements of a structured rescue' (Maynard, 2014). These are:

1. Safety and Scene Assessment
2. Stability and Initial access
3. Glass Management
4. Space Creation
5. Full Access
6. Immobilisation and Extrication

The team approach does not just refer to the team of firefighters at the scene of a road traffic collision alone but to the liaison between the three emergency services (police, fire and ambulance), and potentially others (e.g., highways agency etc.), to reduce entrapment times and, consequentially, mortality rates through better organisation and a methodical approach to the extrication. In relation to the rescue activities that are required by the fire service, it is stated in the National Directorate for Emergency Management (2009: p 28) that the minimum crew on an appliance (fire engine) which would be expected to respond to a Road Traffic Collision would be four crew members, plus the crew commander (i.e., 5 members in total as a minimum). The tasks for a crew of five members can be seen below, excluding the crew commander who has an overall responsibility for scene assessment and coordinating the rescue operation. This view is further asserted by Deliner (2017) in his article *'Vehicle Extrication with Limited Staffing Levels: surviving a cut-back environment'* who argues that a minimum of five fire and rescue staff would be required for a light motor vehicle extrication. Where more crew members or personnel are available, the workload would be shared accordingly to ensure that time critical factors for victim safety are met. It becomes apparent, that where more victims are present, then more resources will be needed to effectively manage the requirements of the incident, with victim safety being the main priority.

Crew Member	Function
Tool Operators (2no.)	Inner circle survey, stability, glass management, operate tools
Driver	Lay out firefighting gear, put out signs, outer circle survey, set up tool staging area, disconnect battery, assist casualty carer and tool operators
Casualty Carer	Assess casualty, deal with immediate needs, keep IC up to date on casualty's condition, keep casualty informed, assist HSE personnel

Figure 3: Allocation of roles at a Road Traffic Collisions

(National Directorate for Emergency Management, 2009: pp 28)

Many fire services from around the world will structure the weight of their initial response on risk-based criteria that will determine the adequate resources that are required to manage anticipated risks and meet the basic demands of the incident that have been identified through their assessment. In the UK, there used to be national guidelines for the weight and timeliness of responses to fire calls. These guidelines were first formulated by the Riverdale Committee in 1936 and had been modified several times, most recently by the Joint Committee on Standards of Fire Cover in 1985 (Home Office, 1998: p - 1). These standards would govern response times as well as pre-determined attendance[24] (PDA) of resources for different types of incidents based on category of risk from A (highest) to D (lowest).

- **Category 'A'**
  Built up areas in large cities containing large commercial and industrial premises or high-rise property where there is a strong chance of fire spread. The recommended minimum first attendance was three pumps, two to attend within five minutes and one within eight minutes.

- **Category 'B'**
  Areas in towns and cities such as smaller industrial areas, extensive shopping centres and factory estates. The recommended minimum first attendance was two pumps, one to arrive within five minutes and the other within eight minutes.

- **Category 'C'**
  Extensive areas of residential dwellings such as terraced property, blocks of flats or light industry/commercial. The recommended minimum first attendance was one pump within eight to ten minutes.

- **Category 'D'**
  Consisting of rural property, villages and farms. The recommended minimum first attendance was one pump within 20 minutes.

These response criteria were abandoned with the introduction of the Fire and Rescue Services Act 2004 that required local fire and rescue authorities to determine their response criteria based on multi-source data acquired as part of their Integrated

---

[24] Pre-Determined Attendance refers to a set number of resources that are despatched in the primary (or subsequent) response to various incidents.

Risk Management Plans that became a requirement under this Act (Dudeney, 2018). All fire and rescue services are since required by a National Framework to develop Integrated Risk Management Plans in collaboration with other agencies, that will assist them in better determining how they profile risk and allocate adequate resources to reduce the commercial, economic and social impact of fires and other emergency incidents in their local fire authority areas (Communities and Local Government, 2008: p 5). This fundamental change shifted the balance from a demand-based funding model to a risk-based funding approach. This has both pros and cons that will be discussed in greater detail later (Fry, 2017).

An example of how West Midlands Fire and Rescue Service determine their responses in accordance with their risk values arising from their IRMP can be seen in Figure 4.

Incident Risk Category	Description	Fire Appliance Attendance Standard
Category 1 High Risk	This category comprises those incidents that present the most significant risk to life, for example house fires, road traffic collisions and water rescue incidents.	5 minutes
Category 2 Medium Risk	The category comprises those incidents where there is a potential life risk (either through incident severity or the type of property affected), but there is a reduced likelihood of this risk being realised	7 minutes
Category 3 Low Risk	This category is made up of those incidents where there is a significantly reduced risk to life, for example the provision of advice only or animal rescues.	10 minutes
Category 4 Secondary Fires	These include fires in the open or fires where houses are not involved, eg fires in fields, gardens, woods, grass, rubbish and skip fires.	20 minutes
Category 5	False alarms	

Figure 4: WFRS response criteria against IRMP

(West Midlands Fire and Rescue Service, 2024: p – 34)

The time of response and weight of resources might vary between different jurisdictions based on determined risk proportionality. Below, is the national PDA for high-rise building fires i.e., any structure taller than 18m (or 6 storeys) for each UK local fire authority. In London, for example, a property of this type will receive a minimum attendance of 5 pumping appliances (fire engines) and an aerial ladder platform (Fire

Brigade Union, 2017) with between 22 – 27 firefighters. The variation in response to high rise premises by local fire authority area is also evident in Figure 5. Conversely, for a single dwelling fire with reports of a fire located to one room, the initial PDA by the LFB might mobilise just 2 appliances (fire engines) with between 8 to 10 firefighters (Dudeney, 2018).

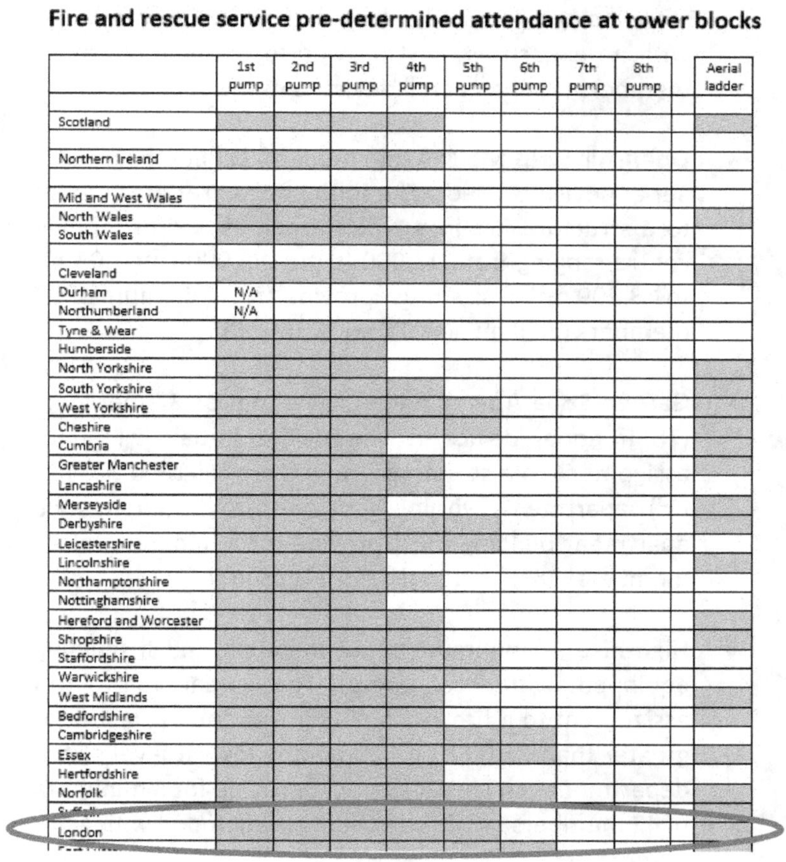

Figure 5 – National PDA response to high-rise (UK)

(Fire Brigades Union: 2017)

In the United States, section 5.2.4 of NFPA 1710 (2016) recommends the following attendance resourcing in staff numbers per property type:

- **Single-Family Dwelling** — minimum of 14 members (15 if aerial device is used). The initial full alarm assignment to a structure fire in a typical 2000 ft^2 (186 m^2), two-storey, single-family dwelling without a basement and with no exposures must provide for a minimum of 14 members (15 if an aerial device is used).

- **Open-Air Strip Mall** — minimum of 27 members (28 if aerial device is used). The initial full alarm assignment to a structure fire in a typical open-air strip shopping centre ranging from 13,000 ft^2 to 196,000 ft^2 (1203 m^2 to 18,209 m^2) in size must provide for a minimum of 27 members (28 if an aerial device is used).

- **Garden-Style Apartment** — minimum of 27 members (28 if aerial device is used). The initial full alarm assignment to a structure fire in a typical 1200 ft^2 (111 m^2) apartment within a three-storey, garden-style apartment building must provide for a minimum of 27 members (28 if an aerial device is used).

- **High-Rise** — minimum of 42 members (43 if building equipped with fire pump) The initial full alarm assignment to a fire in a building with the highest floor greater than 75 ft (23 m) above the lowest level of fire department vehicle access must provide for a minimum of 42 members (43 if the building is equipped with a fire pump).

Standardised responses to incidents aim to ensure that adequate resources will be to hand if needed. Experience, past incident analysis and the requirement for the implementation of a risk-based management approach (i.e., providing adequate controls to manage identified risks) in many countries will determine these minimum numbers, depending on the type and nature of incident. It is worth mentioning that there are also two principal types of incident response systems used by fire services:

a) the re-active conditions-based response, and;
b) the proactive pre-assigned task-based response.

Both of these response systems have advantages and disadvantages. However, in the reactive conditions-based response, there is greater onus on the incident commander to make sound judgements about the resources they need and how they plan to allocate them. It is argued that this method presents higher risk of 'freelancing' in the early stages of an incident (Grimwood, 2008: p 157). The proactive pre-assigned response, on the other hand, makes it easier for incident commanders to identify what tasks crews will be undertaking in order of arrival at an incident. This makes coordination and accountability much easier, although there is a risk that over-commitment to pre-defined tasks when not necessarily needed, where conditions are dangerous or without a robust communication system could have adverse effects on the incident and firefighter safety.

An article in online magazine Fire Engineering (2008) 'Engine Company Assignments', interviews several fire officers with regard to their preference of response models, the majority of which appear to favour the flexibility of the re-active conditions-based response, or as they refer to it, 'Incident Specific Assignments (ISA)'. However, there is evidence to suggest that the pre-assigned task-based response might be favourable for large metropolitan fire services, especially in the USA.

The Fire2000.com research (2005) proposed a Standard Operating Guideline to increase the efficiency, performance and safety of limited crews of three firefighters. However, it is argued, that crews of such numbers will only be able to operate with limited capacity, for it would be physically impossible for the critical tasks to be performed effectively and safely by just three firefighters. In fact, it is identified that three firefighters will only be able to perform 23% of the critical tasks required for a structural fire incident (Grimwood, 2008: p 140). Cushman's (1981) study of the Seattle Fire Department *'Minimum Manning as a Health and Safety Issue'* found that the severity of firefighter injuries declined by 35% when staffing was increased from 3-person crews to 4-person crews. A later study conducted by Joseph Mercieri (2007), evaluated fire appliance staffing level efficiency for Littleton Fire and Rescue in New Hampshire, USA. Based on research conducted by the International City Development Association (ICDA), which concluded that a five-person crew is 100% effective, a four-person crew is 65% effective, and a 3-person crew is only 38% effective, Mercieri, using a linear graph and data calculated by ICDA, concluded that Littleton Fire and Rescue's staffing levels of only two firefighters per shift was 17% efficient in performing interior structural fire attack.

In another case study from the United States, where more than 60 full-scale experiments were conducted to determine the impact of crew size, it was identified that four persons crews operating on low hazard structural fires completed all tasks on the fire ground (on average) seven minutes faster (nearly 30%) than two persons crews, and 5.1 minutes faster (nearly 25%) than three persons crews (Barowy et al, 2010). In April 2010, the National Institute of Standards and Technology (NIST) cited a 1990 National Fire Academy study that found that *"in smaller communities...a crew staffed with four firefighters could perform rescue of potential victims approximately 80% faster than a crew staffed with three firefighters."*

Another study intended to demonstrate the effects of crew size on firefighter effectiveness in high-rise building fires found that: "A five-person crew located the victim on the fire floor 25 minutes and 19 seconds (50.6 percent) faster than a three-person crew and 12 minutes and seven seconds (32.9 percent) faster than a four-person crew". It is understandable that based on the principle of critical tasks that will be required at incidents and the examples given from earlier research, that smaller staffing levels can be considered as ineffective and potentially unsafe.

However, despite compelling evidence, many fire services from around the world are still responding with inadequate numbers of firefighters. Some are doing this is response to budget cuts whilst other fire services do not appear to understand the value of larger crew sizes. Critical task staffing at present will have a level of subjectivity between fire services as the capacity to achieve operational objectives will depend on the variation in crew sizes.

Figure 6 demonstrates the average crew sizes for European fire services. Adjustments might need to be made in order to satisfy the critical tasks per incident type and this will commonly be foreseen in each fire service's SOPs or SOGs. Some fire services, like in Greece for example, do not have a standard for staffing levels. Though it is appreciated that greater numbers will assist in more effectively meeting operational objectives, crew sizes can vary significantly and are commonly low in numbers (2 to 3 firefighters) per fire engine even in the busy metropolitan fire stations.

Country	Required manning level for turnout of a turntable ladder	Required manning level for turnout of a pump water tender	Response time for a fire in urban areas
Belgium	2	From 4 to 6	Turnout within the minute by professionals. Response time 8 to 15 minutes for professionals, and 13 to 20 for volunteers
Croatia	2	4 or 5	15 minutes
Denmark	2	6 in Copenhagen, 4 in smaller towns like Roskilde	Turnout must be within the minute of call acceptance. Response time is 15 minutes
Estonia	2	4	5 minutes in urban areas
Finland	1	4	By region: between 6 and 20 minutes
France	2 or 3	From 6 to 8	Depends on the *département*. Examples: Ain: 20 minutes; Nord: 15 minutes*.
Germany	2 or 3	4 or 5	8 minutes
Italy	1 or 2	5	Depends on the territorial area but must never exceed 20 minutes. In 2009, the response time averaged 15 minutes in central Italy and 13 minutes in the north and south**.
Netherlands	2	6	8 to 10 minutes
Norway	1 in small municipalities, two in towns and cities	At least 3	Turn-out must be within the minute of call acceptance. Response time should be 10 minutes at most in high risk urban areas; 20 minutes in low risk urban areas; and 30 minutes in rural areas.
Slovakia	2 or 3	At least 5	8 minutes
Spain	2 or 3	5 or 6	Varies by region
Sweden	2	5	11 ½ minutes

* The prefects of *départements* tend to lengthen response times to avoid claims.

** Ministero dell'Interno. Dipartimento dei Vigili del fuoco del Soccorso Pubblico e della Difesa civile (2009) *Annuario statistico del corpo nazionale vigili del fuoco*, Roma, p. 54 (*Online*).

Source: ETUI – EPSU survey 2010-2011

Figure 6 – Differences in coordinating of appliances / staffing and response times in EU countries.

If a benchmark for critical tasks for various incident types as described earlier in the research examples could be established, then crew size and consistency would likely acquire greater weight and forethought. However, even countries like the UK that are considered leaders in the fire service industry, are broadly promoting smaller crew sizes and more 'flexible' response configurations to satisfy budget requirements. These smaller configurations are also said to balance the response criteria against actual risk (i.e., why send a big fire engine with 5 firefighters to an incident that might only require 3 firefighters in a smaller vehicle) (Maher et al., 2020). The deployment of fire crews to attend emergency medical incidents has also been seen as an effective solution to utilising available staff that would otherwise be idle at their fire stations (Dearden, 2022). Not all such adaptions are negative, though it does appear that the level of variation in how fire services in the UK justify these alterations is significant, meaning that the same risks are dealt with very differently depending on where in the country the incident takes place. This is also common in other countries.

Cheshire Fire and Rescue Service, in their response profile published on their website show their Rapid Response Vehicle provision. Its primary role is said to be rapid response to road traffic collisions. It carries a maximum of three personnel together with lightweight cutting equipment and rapid intervention first aid equipment for dealing with trauma and critical care on the roadside (Cheshire Fire and Rescue Service, 2023). The justification for this response configuration is quite interesting when compared to the guidance given in the National Directorate for Emergency Management (2009: p - 28) that indicates that the minimum crew on an appliance which would be expected to respond to a road traffic collision would be 5 members in total as a minimum when considering the required tasks for an incident of this type.

Hertfordshire Fire and Rescue Service (HFRS) in their report, SMART Firefighting (2017: pp 16 - 17) point out that the concept of rapid response is not new but responding in a specifically designed fast response vehicle is a different approach for fire and rescue services to consider contemporarily. The report further mentions that 'rapid response capability' is designed to make a rapid intervention in order to either bring an incident to a successful conclusion in its earliest stage, or to provide an early first attack while putting in place a structure that will enable oncoming resources to be offensively deployed quicker than they would otherwise have been. However, it is argued that rapid response alone does not by virtue reflect a risk proportionate response as would be seen in 'Tactical Response' configurations. This latter type of response only deploys in Tactical Response Vehicles (TRV) to specific categories of calls, predominantly secondary fires[25], fire-alarm calls and low-level special services.

There are several fire services identified throughout the UK that have TRV capability in order to help rationalise their response configurations and maintain the resilience in terms of operational capability. As such, it is identified that the TRV concept is simply a means of providing a more appropriate response to more low-level incidents. Conversely, West Midlands Fire and Rescue Service utilise a Brigade Response Vehicle (BRV) as an initial first attack capability for all incident types (with the exception of Hazmat incidents). Crewed with three firefighters, the vehicle is equipped to provide an early intervention at all incidents ranging from fires through to road traffic collisions. It is argued that the key point to understand is that this capability is not designed to bring the most serious or significant incidents to a conclusion but to provide an early first attack and put the building blocks in place for oncoming resources.

In a recent article published by ITV online news (2019), Steve Healey, the Chief Fire Officer of Cumbria discusses the benefits and justification for the use of Rapid Response Vehicles. He states that

> *"New technology and procedures allow these vehicles to be crewed by fewer staff than a standard fire engine and will increase the availability of our resources in areas of relatively*

---

[25] Secondary fires are generally small outdoor fires, not involving people or property. These include refuse fires, grassland fires and fires in derelict buildings or vehicles, unless these fires involved casualties or rescues, or 5 or more pumping appliances attended, in which case they become primary fires.

> *low risk or where there is another standard fire appliance at a fire station close by. They are equipped to be able to respond to, and deal with small incidents on their own and can also attend more serious incidents, such as house fires and road traffic collisions, alongside standard fire engines, and other resources. The obvious benefit the RRVs have are that they are far more flexible and agile than full sized appliances. They are all-terrain and given some of the challenges we have in rural areas and with severe weather events that can make areas hard to reach, these vehicles will be invaluable in providing immediate intervention to saves lives and prevent escalation of incidents."*

In response to this, Cumbria's Fire Brigade Union Secretary, Ed Borrows (2019) argued that:

> *"We believe without the austerity we are facing; this would never have come to the table. Reduced crews, equipment and water will lead to firefighter and public safety being put at risk. No one can predict a serious fire; the public deserves the right resources with adequate firefighters to deal with the incident efficiently and effectively".*

An earlier Fire Brigades Union article (2017) reveals how fire crew sizes 'for a fleet of controversial small fire vehicles in North Yorkshire' will increase from three to four firefighters as a result of pressure from the Fire Brigades Union. They argue that this move could save up to 20 firefighter posts and better safeguard the public and firefighters.

This debate is not restricted to the UK alone and several online forums and articles were identified discussing the contention around smaller crew sizes and the use of rapid response vehicles with both supporters and oppositions alike. Retired U.S. Battalion Chief, Robert Avsec, in his article 'Rapid response vehicles: The answer to downsizing fire apparatus' (2017) gives three key reasons why fire departments should look to downsize their fleet. He says that *"smaller career fire departments and their volunteer department brethren should consider replacing their existing type I or type II engines with rapid response vehicles for some of the following reasons"*:

- Those engines are a poor return-on-investment for the communities they serve. With new pumping apparatus costing $500K or more, is it really a sound investment to purchase a piece of pumping apparatus with a 1,000-GPM pump and a 1,000-gallon water tank when most of the fires the department responds to could be extinguished with 100 gallons of water or less?

- With the personnel turnover rates of smaller departments (e.g., when employees leave for a position with a larger department for more pay) and the retention problems faced by volunteer departments), is it worth the risk of having inexperienced or poorly trained operators driving today's massive fire pumpers?

- Our infrastructure of roads and bridges in the U.S. is in a terrible state of disrepair and it's not likely to get better anytime soon. How many roads and bridges in smaller communities and rural areas can withstand fire apparatus weighing 24 tons or more?

Avsec also discusses the significant advancements in fire suppression technology with the use of Compressed Air Foam Systems, the technical qualities of which make it far more efficient than water in much smaller amounts. He discussed the use of battery-operated rescue equipment that can be easily deployed, is lighter in weight and does not require so many components or fuel to operate whilst possessing the same technical capabilities as traditional fuel driven equipment. Avsec believes that it would be archaic to not perceive the benefits in these developments and adapt to using or implementing such systems in order to improve efficiency.

Conversely, there is plenty of research to demonstrate that crew size is critical to the successful outcome of an incident. It promotes safer working conditions for firefighters and, as a result, they are more efficient and effective at protecting the public. Crew size requirements have been assessed against the critical tasks required to effectively manage specific types of incidents. Thus, there is a direct association between the number of staff required to address the critical tasks at any given incident. The more the critical tasks, the more staff will be required. Unions and other sector professionals alike have been vocal at pointing out these associations. However, despite the evidence, many fire services from around the world continue to search for innovative ways to reduce crew size and response configurations in order to address budget cuts in attempt to make their fire service more financially efficient by proportionately matching response to risk. Data analysis has been fundamental in providing evidence to justify these changes.

The implementation of smaller, cost efficient and risk proportionate response configurations could be utilised effectively within a specific context. Although it might seemingly be ideal to have traditional responses consisting of large fire engines with lots of firefighters attending every incident, formal metrics are demonstrating that this is neither needed nor financially viable. Problems arise where fire services are utilising low crew numbers by default. Where there is an absence of organisational standards and risk assessment parameters to govern crew size configurations, it is also likely that other operational procedures will not be in place either. Standard

Operating Procedures and Guidelines are designed against a series of risk critical factors and incident specific tasks, and establishing crew capacity is fundamental to effectively, efficiently and safely meet incident objectives.

Standard Operating Procedures and Guidelines

It is acknowledged by Fire Service expert and Associate Professor at Lund University (Sweden), Dr Stefan Svensson, that, operational procedures are, to an extent, based almost exclusively on experience, implying that this does not undermine their functionality "at least in most cases". Firefighters learn and develop through experience and acquire fluency through exposure, trial and error, and through association and repetition. During discussions with veteran firefighters, they often refer to a method of extinguishing fires as "putting the wet stuff on the hot stuff". Though it is acknowledged that this is a greatly simplified explanation, it is indeed the foundation to fire extinguishment. In order to consolidate this knowledge into the organisation's operational standards, it must somehow be systematised through a formalised process to be efficient. As argued by retired Fire Chief, Thomas Warren (2016), *'for any emergency response to operate efficiently, a formal approach to the incident must be taken. Efficient fire department operations will eliminate duplication of effort, increase accountability, and provide a higher level of situational awareness. Every firefighter on the scene knows what is expected of them as well as each of the other companies (units) operating on the emergency scene'*.

Generally speaking, Standard Operating Procedures are derived from a mix of experience, lessons-learned from previous incidents and, more so nowadays, an inclination to provide validation through empirical scientific evidence achieved through extensive and approved research methods (Benny, 2019). Occupational Health and Safety laws have also contributed significantly to shaping current operational guidelines. Having these processes documented serves several purposes. Procedures form the basis from which firefighters operate to achieve specific objectives. They are performance based; identifying critical tasks that must be undertaken at an incident and help create a systematic approach to ensure that mission objectives are achieved in a safe and effective manner (Butler et al., 2021).

A less obvious benefit is that they limit the risk of personal liability by discharging responsibility onto the organisation where failures might occur, as these failures will, by virtue, be organisational if staff have demonstrated commitment to following operational protocols. Where staff cannot demonstrate adherence to these though, there will be clearer lines of accountability. Furthermore, documented procedures assist in determining and identifying points of error, and lessons learned from these will effect change in the organisation through their policies and practices.

The term and use of Standard Operating Procedures in recent years has changed for the more favourable term 'Standard Operating Guidelines' (SOGs). The difference between these two terms is that a 'procedure' infers a level of rigidity that is not commensurate to the dynamic nature of fire service operations. As such, since emergency incidents are unpredictable and flexibility is essential, the use of 'guidelines' now appears to be the more desirable term. It is argued that guidelines are believed to reduce the need to identify exceptions, and perhaps even limit liability due to actions by firefighters (FEMA, 1999: p 2). The language used for SOGs is different and allows more flexibility at the emergency scene than procedures, that often dictate actions in a policy-like format using definitive terms such as 'you must', 'you shall' and so on (Murphy, 2008). However, there is a school of thought that argues that the SOP versus SOG argument in fact boils down to a matter of semantics and that there is no difference in terms of liability when coming to legal proceedings as a result of an injury or fatality (Comstock, 2009). Though there is some contention around the intricate meanings of 'procedure' and 'guidelines', these terms will be used interchangeably throughout This book to describe a structured system of guidance to achieve operational objectives at fire service incidents.

It was demonstrated earlier that procedures are not always effective in overcoming habitual behaviours that are embedded in fire service culture (Gough cited in Grimwood, 2017: pp 442 - 445). This is either due to the persistence (or normalisation) of such behaviours, the dynamic that they represent within their cultural context or simply because procedures are not appropriate for the circumstances, or not enforced. This is indeed a challenge for many fire services and a major contributing factor in firefighter injuries and fatalities. However, without procedures or guidelines to inform fire service operations and govern how firefighters operate, the likelihood of what is known as 'freelancing' will be significantly increased – and with freelancing the risk to fire fighter safety and ultimately to the outcome of an emergency incident intervention increases exponentially.

The Problem with Freelancing

Where deviations from standard operating procedures occur or where there is no appropriate foundational basis to firefighter training, then freelancing is more likely to occur. This is a logical consequence as, if there are no standards to meet or conform to, then there can be no integral structure or control of the firefighter actions. Freelancing can be seen as an unstructured and uncoordinated method to meeting operational objectives. Van der Feyst (2013), a firefighter and training instructor from Ontario Canada, describes this as follows; *"Freelancing, by definition, is when individuals work independently, commit to tasks and acts without the expressed knowledge or consent of an officer or incident commander. Essentially, it is when firefighters do what they want to do, when they want to do it".*

It can be argued that freelancing does not apply solely to individual firefighters but can be representative of a unit or team of firefighters, irrespective of rank, if they are operating independently of each other and not to a specific operational procedure or guideline. Based on this assumption, entire fire service operations, even at large scale incidents, can be structured around freelancing. Such cases usually lead to what is referred to as the 'chain of errors.' This principle is based on the concept that rarely one mistake alone will lead to a catastrophic failure (Price, 2010: p - 189) and, as such, there needs to be a chain of events that create a domino effect that will in turn govern the negative outcome of that incident. There are 10 critical indicators identified in Paul Grimwood's 'Euro-firefighter' (2008: pp 26-27) that contribute to the error chain. These are separated into operational factors and human behavioural factors and eight of these are directly associated to patterns in freelancing behaviour:

- A failure to meet competency.
- A failure to meet tactical objectives.
- Use of an undocumented/unauthorised procedure.
- Departure from Standard Operating Guidelines.
- Violating limitations.
- Inadequate leadership.
- Tunnel vision, and;
- Inadequate or inappropriate communication.

Alan Brunacini was a retired Chief of the Phoenix (Arizona) Fire Department before passing away in 2017. He is referred to as an icon of the fire service world due to his extensive contributions and authorships to the field for over 50 years, especially around firefighter safety. In his article, 'Freelancing isn't Free' (2010) he talks about the importance of establishing command once arriving at an incident and how establishing control in the initial stages is fundamental to the outcome. He makes reference to a logical objective of incident command that states *"start under control, remain under control, never lose control'"*. Brunacini argues that freelancing is counter-intuitive to this objective and is governed by *"our strong and historic obligation to act along with the current need to act"*. This, he argues, *"can create a natural initial conflict between action and command"*. Without an agreed-on game plan [i.e., standard operating procedure (SOP)], this action / command conflict can create a huge initial arrival challenge. He continues by stating that *"if we let the urgency of the event cause us to completely skip command and instantly attack the fire, we are engaging in what is commonly called 'freelancing'"*.

It is worth making reference to Brunacini's insights into the dangers of this common practice known as freelancing:

> *"All this "take command first" talk comes with its own built-in dilemma. .... the incident requires fast, effective action, and we responded to do exactly that. The dilemma is that the initial officer who skips command and goes directly for the actions is buying a ticket in the "freelancing lottery." This lottery works opposite from most other lotteries, where you pretty much always lose. In the freelancing lottery, you generally win. Simply, most of the time freelancing puts the fire out, everything gets better, no serious safety problems occur, we high-five each other in the front yard, and we keep wearing off-duty T-shirts emblazoned with suicide slogans. Winning just naturally makes you want to play*

more. Freelancing generally works: It takes advantage of the skill and spirit of our firefighters, it is very (!) fast, it puts water directly on the fire, it runs past / over anything or anybody who tries to slow it down, and it has no time or patience for anything bureaucratic (like command). It is the fastest, shortest, and most energetic response we can extend to an urgent problem. Firefighters are smart. They do what worked last time. That is the reason we keep freelancing incident front ends. The problem happens when, after getting 152 winning (or close-call) freelancing tickets, we come up with freelancing ticket number 153, and it is a loser. What a losing freelancing ticket means is that the past 152 wins seduced you into a fatal situation and that now your buddies are going to get to wear their dress uniforms and will hear some really sad bagpipe music in three days. At that awful moment, freelancing is no longer free."

Freelancing is a globally recognised problem and is often referred to as something that is 'outright dangerous' that fire services should strive to control through effective command and control structures, policies and procedures (Van der Feyst, 2013). The problem with freelancing is that it is unpredictable, and this comes as a significant shortfall when so much of what firefighters do requires coordination and teamwork. Fire service operations at times can be very dangerous. The level of risk is also increased when things happen that firefighters don't expect will happen. These are not necessarily unforeseeable risks, a lot of the time they are simply ignored because nothing bad or dangerous has happened at similar incidents attended in the past (Gough, 2018). Complacency, routine and the lack of guidelines contribute to freelancing and these present significant risks to firefighters and members of the public alike (Reason, 2017). Yet even today, it has been identified through this study that there are fire services around the world where freelancing is common practice and broadly accepted - even at organisation level across all ranks.

It appears from this research that fire services in northern-Europe that have a structured systems-based approach when compared to south European fire services are generally better at addressing issues around freelancing. It is not an acceptable practice and firefighters are less inclined to be put at risk by freelancing colleagues. Procedures also assist in controlling freelancing opportunities as a deviation from accepted practices will not be received without accountability or repercussions at an organisational level. It is also easier to identify where such practices are occurring due to the easily identifiable deviations from the SOPs. Robust command and control structures will further ensure that freelancing is kept in check. This is iterated by Steve Dudeney, a senior officer with the London Fire Brigade (since retired) in an interview in 2018 where he states that even in the early days of command and control, efforts were made to limit the likelihood of freelancing. He states that *"in the early days of command, command was very similar to military leadership. You would have one person with one plan and in essence that still works today. Whoever the incident commander is, you work to that plan, you don't have freelancing because how can you take control of a fire if you've got 30 or 40 or 60 individuals all doing their own thing, or even teams of individuals you know, six, seven, eight teams doing their own thing".*

Additionally, freelancing is also regulated through training, as this will provide the foundations for putting policies and procedures into practice. Training teaches a specific methodology or process for achieving certain goals that is perceived as correct (or accepted) therefore, freelancing is often discouraged where training is appropriate and consistent. Despite the robust systems employed by many world-class fire services, relevant literature would suggest that freelancing is still an issue of concern. If this is the case for fire services with robust structures and systems of work, then there is little confidence that freelancing will be dealt with effectively in fire services that do not possess the same ethos around the development and application of defined processes, policies and procedures.

Firefighter Competency and Training

Fire service training and the requirements for competency criteria are a continuously evolving process governed by several variables. In reflecting on this evolution, Grimwood (2008: pp 2-3) describes how *"things are done a lot differently in the fire service these days"*. He then goes on to explain *"when I joined in 1971, we had plastic gloves, turnout trousers and a smart woollen fire tunic. We would tackle most fires without the protection of self-contained breathing apparatus, as this was considered as a sign of weakness amongst our brothers"*.

Such practices were also described in Clifford Thompson's 'Falling Though Fire' (2017: p 92) where at a fire at a derelict cold store in Charterhouse Street in the City of London in the late 1980s he describes how he, under the instruction of a more senior officer, embarked on some old-fashioned firefighting commonly referred to as 'smoke eating', where access was made deep inside a smoke-filled structure with no breathing apparatus. Many more examples are described in both these publications, but also in others from around the world by current and ex-fire service personnel, describing antics that would simply be considered as unacceptable in their respective fire services today. However, these anecdotal events are not discussed with the intention of glorifying their actions, but contrarily, to demonstrate the extent of ignorance, or perhaps the machismo culture, of that era and how procedure and practice has progressed since then.

In the 'Good Practice Guide for Fire and Rescue Services' of the National Operational Guidance Program (2016: p 4) the importance of learning is referred to as follows: *"Knowledge learned from operational activity should be shared with all fire and rescue services, and the wider sector where appropriate, to help reduce risk. A consistent and shareable learning process will enable fire and rescue services to improve performance at incidents, cope with future changes in the operational environment, develop and innovate to provide the best service to the public and improve the safety of firefighters".* Firefighter competency plays a critical role in assuring that fire services meet their key objectives of protecting life and property in a safe and professional manner. However, apart from learning through sharing processes (i.e. the experience-based approach), it is also essential to keep pace with the advancements in science, technology and trends of modern society that are introducing new and complex challenges.

When asked 'how important science and engineering principles are in shaping operational procedures to ensure an effective and resilient fire service', Professor Stefan Svensson (2018) responded that '*if we are able to have a sound scientific basis in our education and training of firefighters and fire officers, we should be able to shape operational procedures into something effective and resilient. It's not an either/or. It's about science and experience in harmony. Neither of them is better than the other. We need all of it*'. Nonetheless, the reference made above by the National Operational Guidance Program perhaps represents an ethos that is not widely, or readily, shared by all fire services and would appear from this study to be culture specific. Undoubtedly, the examples given by Grimwood (2008), and Thompson (2017), are still prevalent today in some (perhaps many) fire services around the world. There are numerous examples on the internet, on websites, forums and videos that demonstrate examples of poor firefighter performance and practice at live incidents. By typing the phrase 'firefighter fails' into a search engine the author accessed hours of footage of fire-firefighter accidents or performance failures at incidents and training, the vast majority of which are resultant of a poor culture around training and competency criteria or a failure to adhere to operational procedures.

It could further be argued that if, for example, firefighting tactics and training of a specific fire service are outdated then what would the point be in sharing information and experiences as this will only recycle poor or useless information, feeding into a vicious cycle of bad practice. And if this is indeed the case, will critical elements that would have the potential to ensure progress and learning be identified in the first place? As argued by Thompson (2014) in his article 'Training Basics and Essentials for the Fire Service', *"when we fight a fire from which nothing is learned, that fire is merely an event. Training that does not connect to reality is useless"*. Every fire service should have a minimum standard of training that reflects the requirements of the organisation in being able to meet its operational objectives. To cope with the requirements of X incidents a defined Y level of training must be achieved and maintained as a minimum by all members of staff. Anything outside this standard would be considered as unacceptable. It makes sense that the acquisition of a set level of competence fundamentally forms the basis (and pre-requisite) for employment in any such service.

Training is a cyclic process structured around identified needs (performance requirements) (Devi & Shaik, 2020). Needs are established through valid methods of data analysis scientific research, operational experience and evaluation of previous incidents from where traditional approaches evolve into more effective methods (Madrzykowski, 2010). The outcomes drawn from this methodological approach will determine the minimum level of competency criteria that is required and then training should be repeated periodically to ensure maintenance of competency standards along with the implementation of new developments to training and so on. As such, it would be expected that firefighters will be competent to undertake specific key tasks expected of them in the course of their duties.

In an interview with Steve Dudeney (2018) he argues that even as trainees, appropriate guidance should be available through the incident command structure and the assistance from more experienced firefighters Control measures and process mapping should be in place in the form of guidelines that will assist in establishing those key roles and functions at a given incident. He further states that in the UK, this is done through National Occupational Standards which has set National Competency Criteria for firefighters that should be achieved by every serving operational firefighter and maintained through a consistent and periodic training regime.

In the United States, there are the NFPA standards, amongst others, NFPA 1001 (2019) for example, that identifies the minimum job performance requirements (JPRs) for career and volunteer firefighters. However, as aforementioned these standards are not universally applied across the United States with many services maintaining their own training standards. These standards are, in the most, considered as best practice guidance, and in some other cases, Federal or state compliant Occupational Safety and Health Administration agencies have incorporated wording from NFPA standards into regulations for firefighters. However, it is still very much the case that many fire departments in the USA will apply these standards voluntarily and adjust them to fit their own systems of work.

Some departments will utilise their own standards which creates issues in terms of interoperability between different fire services working at the same incident (Spell, 2022). The Fire Department of New York (FDNY), for example, generally implements its own standards and procedures, the New York

City Fire Code, that are not affiliated with NFPA codes of practice (NYC Bureau of Fire Prevention, 2021)[26]. As one of the busiest fire departments in the world with a strong legacy and wealth of experience, FDNY's procedures are tried, tested and adapted to the diverse and complicated nature of incidents they attend.

The Operational Competence Guidance (OCG), developed in 2013 provides UK Fire and Rescue Services (UK FRS) with a competence framework for firefighters and commanders to ensure that they can safely and effectively apply the necessary skills, knowledge and understanding in an operational environment (CFOA, 2013: p - 3). Additionally, National Occupational Standards (NOS) provide a structure that enables fire and rescue services to design and deliver a consistent level of training to a common standard, across all levels of command. NOS is structured around the principles of the Incident Command System and broadly consists of eight separate units across all four levels of the command structure (CFOA, 2013: p – 5).

It is stated that National Operational Guidance (NOG) is intended to be the foundation for developing operational policies, procedures and training for firefighters to deal with incidents effectively and safely. It is industry 'good practice' for everybody in fire and rescue services to draw upon. The process driven system implemented in the UK, at least at first glance, appears to provide a reasonable benchmark for other fire services. The independent performance audit process conducted by His Majesty's Inspectorate of Constabularies and Fire & Rescue Services (HMICFRS) provides added value to this structure as it aims to ensure that organisational objectives are being satisfied as planned providing a valuable fire service performance metric.

---

[26] Although it is suggested that some elements of NFPA standards are incorporated into FDNY codes.

The fundamental values of training standardisation are acknowledged by Adrian Ridder, at the time of writing, a Fire Safety Engineering student at the Bergische Universität Wuppertal in Germany, who as part of his final year was required to spend a mandatory term of 12 weeks in a practical working Fire Safety Engineering environment. His comparative study of *'Fire Services in Germany, Queensland & New South - Experiences from 4 months in Australia'* (2011) identified that in Germany, every city and even small town had its own independent fire brigade, funded by the city. This has led to a huge number of approximately 25,000 brigades in all of Germany with significant differences in training, equipment and tactics, though national standards on basic equipment such as hoses and radios ensure a certain level of interoperability.

In Australia, training standards are not nationalised but adjusted to meet each fire services' needs. Though, in principle, there are commonalities in the fundamental elements of the training curriculum, the functional SOPs might be more fire service specific. Ridder (2011: pp 6-7) concluded that the state-wide model adopted by Queensland Fire and Rescue Service was far superior to that of New South Wales in delivering its services, as the rural fire service in Queensland was included within the overarching organisational structure, the division of responsibilities between the different emergency services seemed to be in order. In contrast to this, the relationships between the full-time and volunteer fire services in New South Wales were described as fraught with infighting and mistrust. It is implied that this lack of standardisation resulted in inefficiencies and inconsistencies in service delivery.

In order to ensure the maintenance of continuous and cyclic process of training, there must be commitment reflected through organisational leadership and culture which is geared around understanding both the experienced-based approach to learning as well as keeping pace with advancements in science and technology and how these might impact on service delivery if not captured through training. One can look at the technological factors that would broadly influence firefighting operations.

In Volume 2 of his trilogy on fire behaviour, 'Fire Dynamics for Firefighters', Ben Walker (2017: pp 20 - 21) discusses how modern furnishings and equipment such as mobile phones, laptops, tablets etc. now have a higher energy potential, facilitating hotter and faster burning rates than more traditional materials, creating greater complexities for firefighters. This, in conjunction with modern methods of building design, creating more insulation within compartments that does not allow heat to dissipate effectively, can lead to a greater likelihood of rapid-fire development phenomena (flashover or backdraft etc.).

In attestation to this, the New Science Fire Safety Journal (2012: p 12) found that the onset of flashover[27] is 8 times faster in a modern furnished room, reaching this critical point in under five minutes than in a room of traditional materials of 50 years ago where the average time to flashover was 29 minutes. These figures might be slightly overestimated as another report by the Swedish Rescue Service Agency (SRSA) identified that in the 1950s, the average time from ignition of a fire to flashover was 15 minutes and by the 1980s this was down to under 5 minutes (Grimwood, 2017).

Further to this, it can now take only two to three minutes for potentially fatal conditions to occur due to the increased amounts and toxicity of the smoke resulting from the more frequent use of plastics in building contents (Harrami and

---

[27] Flashover is a natural progression of the fire whereby a fire within a compartment transitions from the development stage into a fully developed stage, consuming all combustible products within the compartment. It is the stage of highest heat release rate energy and often referred as the point at which a fire in a room becomes a room on fire.

MacIntyre, 2006: p 4). In order to fully understand this fundamental difference, Harrami and MacIntyre (2006: p 45) further describe how a hundred years ago the fire brigades had special groups of personnel who were responsible for salvaging belongings from burning buildings. They could operate this way because they knew they had 15-20 minutes in which to take out possessions. Photographs from the 1950s and 1960s show belongings placed in the garden at a safe distance from burning buildings. This is almost never seen nowadays.

In respect of these changing circumstances, Grimwood (2017: p 1) argued that:

> *"...one thing is certain, that the basic principles of firefighting intervention must further evolve, for failing to do so may expose firefighters to increased hazards. Just as the dynamics of compartment fires have changed over recent years due to the increasing contribution of plastics in the fire load coupled with an increasing energy efficiency incorporated in the boundary walls, lightweight buildings are becoming less stable under fire attack than the traditional constructions of past experience, so we must make sure that our tactical approaches ensure the working environment is understood and any unnecessary exposure of risk to firefighters is countered".*

Indeed, such developments will need to be reflected in the training structure of fire services if they are to be able to understand and deal with the impact of these changes effectively – and, to a great extent, many fire services are doing this – even if this is achieved through trial and error, or the proverbial 'experience-based' approach.

In relation to training, taking Compartment Fire Behaviour Training (CFBT) for example, extensive research has demonstrated that different materials commonly found within structures (residential and commercial) have an estimated heat release rate when they burn. This is the rate at which fire releases energy - also known as power - and is commonly measured in Watts (W)[28], Kilowatts (KW) or Megawatts (MW). Some indications of these heat release rates for various materials can be seen in Table 3:

ITEM	HEAT RELEASE RATE
Sofa (2 seat)	3.0MW
Sofa (3 seat)	3.5MW
Upholstered Chair	2.0MW
Small Rubbish Bin	0.3MW
Single Mattress	1.0MW
Pine Bunk Bed	4.5MW
3-side Workstation	7.0MW

Table 3: Heat Release rate of common household furniture

(Walker, B. 2016: p 72)

Conversely, water has a specific heat absorption capacity when transitioning from a fluid into steam (vapour). Steam then has a further important role to play by displacing oxygen and mixing with fire gases (pyrolysis gases), reducing their flammability levels and making them inert. Therefore, in order to achieve effective control of a fire, the amount of water must be capable of absorbing the heat energy released by the fire via the production of steam, and the amount of steam produced must be capable of displacing oxygen and neutralising the fire gases.

[28] A Watt is equal to one joule per second or J/s.

This can be demonstrated by the following mathematical calculation:

- A well-furnished living-room of 4m (w) x 5m (l) x 2.5m (h) occupies an area of 50m^3. With normal furniture and appliances expected to be found in a common living-room it is anticipated that a fire in this room would have a heat release rate of around 5MW.

- A firefighting hose of 230 litres per minute is used to tackle this fire.

- 230lt/min = 3.83lt/s (or 3.83kg/s)

- 3MJ/kg is the default value for the amount of energy that one litre of water can absorb.

- 3MJ/kg x 3.83kg/s = 11.49 MJ/s

- 11.49 MW

- 2.61 m^3 is the default quantity of steam produced at 300^{0}C per litre of applied water, therefore,

- 2.61m^3 x 3.83lt/s = 10m^3/s

This means that the water from the firefighting hose would need to be applied at 230 lt/min for 5 seconds (for a room of 50m^3) in order to acquire the required amounts of steam to satisfy, or assist, the extinguishing process. As the intrinsic techniques and effectiveness of water application is within the region of 50% in a best case scenario and generally around 25% for the average firefighter, this means that an additional 75% compensation should be factored as a minimum flow-rate capability in a structural fire. This would bring the minimum flow-rate capacity to around 400lt/min.

(McDonough, 2018)

The above is only a simplified analysis of this process. However, it is sufficient in demonstrating that simple mathematical calculations based on the principles of physics provide an empirical basis on which to found elements of operational tactics and training. It could be argued, that based on the scientific elements of Fire Behaviour Training and the physics of fire, it should not really matter whether one is in New York, Venezuela or the Congo, the way in which a fire will develop inside a compartment will be relatively predictable (once accounting for specific variables such as the type, size and height of the structure, the available ventilation opening and the materials therein.), or at least set the general principles around firefighting tactics. Therefore, theoretically, any standards to achieve an effective, systematic and safe method of dealing with such incidents.

Similarly, if one were to take the example of Road Traffic Collisions (RTCs), there are several systematic protocols in place for these, like SAVER (Systematic Approach to Vital Emergency Response), the Road Traffic Accident Handbook (2009) etc. The general provision of any such protocol is that it focuses on meeting victim critical requirements and sets a series of priorities that should be considered and then followed when attending such incidents. With the proliferation of electric vehicles and advancements in automotive technology, both firefighting and rescue operations will face significant new challenges that if not captured through training, will impact on the capacity of firefighters in responding to such incidents effectively and safely (IAFC, 2021).

It has been established that training and competency are essential to the effectiveness of fire service operations. Training must be structured against incident specific needs with the of intention of automatising the employment of critical tasks by on-scene firefighters. Automatised tasks will allow for quicker and more efficient actions at emergency incidents and will create monitorable guidelines to ensure that firefighters are kept safe and doing what they are meant to be doing i.e. not freelancing. Procedures and guidelines should not only consider past experience as the main driver for their development, as this alone does not form a reliable metric. Experience-based training is often focused on outcome, and a positive outcome does not necessarily mean that these methods are correct or most effective. Therefore, training must also incorporate other scientific and evidential bases to structure procedures and guidelines (as well as the impact of new technologies on fire service operations) and, in turn, the training requirements for firefighters. In addition, both procedures and training can only be effective if staffing levels in fire engines (or other fire and rescue emergency vehicles) are standardised.

Training for procedures with a 5-person crew can only be as effective if the fire engines are staffed with 5 people or a contingency exists when lower staffing numbers occur. It has been identified in the literature review that training between fire services on a global scale varies significantly. This was also apparent in the sample groups of north European and south European firefighters. This is not necessarily a problem providing that a series of minimum standards are set that will consider incident critical tasks, crew size, minimum competency criteria and current technologies, practices and research. This did not appear to be the case in the study group of north and south European firefighters.

The Integral Role of Incident Commander

It would generally be expected that the Incident Commander, once at the scene of a fire incident, would firstly conduct a dynamic risk assessment of life risks, the extent of fire (location, possibility of spread and contributing factors) and other hazards before allocating tasks to crews (Dudeney, 2018). This would require a 360° assessment of the incident area – where this is possible. It is also seen as essential for the Incident Commander to not engage in firefighting or other hands-on operations so that they are able to effectively fulfil their coordinating role.

The London Fire Brigade's Incident Command Procedures Policy[29], number 238 (rev 2017), summarises the Incidents Commander's role and responsibilities as follows:

The Incident Commander arriving as part of a pre-determined attendance must:

- Conduct a dynamic risk assessment of all aspects likely to affect the overall command of the incident and declare the appropriate tactical mode[30].

- Ensure that any tendency for crews to commence instinctive action is held in-check until positive directions are given.
- Ensure at all times that actions being taken complement the primary objectives of operational incident command including to save life, prevent further spread/ loss/risk of injury, mitigate damage, mitigate

---

[29] These procedures / policies fall in line with National Operational Guidance for the UK fire service.

[30] The Tactical Mode is distinguished into two categories; Tactical Mode Oscar - TMO (Oscar being the letter from the phonetic alphabet for Offensive) that would indicate the implementation of an offensive strategy for dealing with the incident, as with Tactical Mode Delta – TMD (Delta being the letter from the phonetic alphabet for Defensive) would indicate a defensive strategy.

environmental damage and minimise disruption to the community.

- Will use the Decision-Making Model (DMM) when handing or taking over an incident or briefing others to ensure that all relevant information is exchanged.

The Decision-Making Model is a cyclic process to be considered by the Incident Commander and will incorporate elements such as:

- An assessment of the situation to establish priorities and decide upon an overall action plan.
- An assessment of additional resources required. It is stated here that it is preferable to over-estimate the resources required at the initial stages of the incident.
- The designation of specific duties including the setting up of the appropriate level of command support.
- Develop an effective organisational incident command structure and communications network.
- Decision of resource needs and requesting reinforcements.
- The allocation of resources to sectors or specific tasks.
- Receiving and acting upon incoming information, revising and developing the operational plan.
- Ensuring that all necessary actions and investigations in the event of legislative infringements, suspicious circumstances, injuries to personnel are immediately undertaken.
- Requesting additional and specialist personnel / assistance in order to undertake such duties.

Fire services that have standard staffing levels on their front line appliances are able to implement such a structures more easily. This is, of course, not the case for all fire and rescue services. Some fire services have down-sized the minimum number of firefighters permitted to attend incidents down to 3 firefighters per appliance whilst in some countries staffing levels might only consist of two firefighters as identified in the research data, mainly for south European fire services. This means that the integral role of incident command at least at the initial response stage can be compromised as it will be necessary for officers to engage in hands-on activities as a result of the low staffing numbers.

During the International Fire Instructors Workshop in June 2019, one of the non-negotiables discussed was that the Officer in Charge (OIC) should conduct a 360° assessment of the incident where possible. If further resources are required beyond the initial response, then the OIC's role will sectorised to incorporate more officers and gradually more senior officers depending on the severity of the incident. Initial OICs will hand-over command to more senior OICs upon their arrival and, existing officers will be allocated to different sectors of the incident who will have the responsibility of manging a defined number of resources (or tasks). This process is broadly associated to the Incident Command System that was developed in the United States in 1970s to remedy a series of critical failures in incident command following a series of disastrous forest-fires in California (Stumpf, 1999: pp 1–2). The principal issues that were identified were that:

- There were too many people reporting to one supervisor.
- There were many different emergency response organizational structures.
- There was a lack of reliable incident information.
- There were inadequate and incompatible communications.
- A lack of structure for coordinated planning between agencies
- Unclear lines of authority.
- Terminology differences between agencies.

- Unclear or unspecified incident objectives.

The ICS was designed to assist emergency services by developing a standardized approach to the command, control, and coordination of on-scene incident management. It specifies an organizational structure for incident management that integrates and coordinates a combination of procedures, personnel, equipment, facilities, and communications (FEMA, 2018). Many fire services from around the world either adopt the ICS or another equivalent incident management system. For fire and rescue services, the ICS provides a structure that ensures a competent person is responsible for command and control at operational, tactical and strategic levels. Personnel, sectors and functions should be appropriately supervised to achieve the incident plan.

The incident commander is the nominated competent and responsible person and as aforementioned, the extent of incident command responsibilities will be directly associated with incident severity and time factors, meaning that even low ranking staff might hold command responsibilities upon arrival at an incident. OICs can delegate some responsibilities to others; however, they remain accountable for health and safety at an incident[31]. This point is essential to understand as responsibility and accountability are significantly compromised where low staffing numbers are present or when a standardised process is not followed. When a 3-person crew is in attendance at an incident, the driver must operate the pump and the Officer in Charge (OIC) is responsible for conducting a 360° assessment of the incident area, gathering and communicating relevant information, conducting a dynamic risk assessment, evaluating the required resources and setting operational priorities. This

[31] https://nfcc.org.uk/control-measure/control-measure-command-roles-and-responsibilities/

leaves one firefighter available to deploy hose lines, perform rescues if there is an immediate life risk, undertake firefighting.

Such a crew configuration at a residential fire where interior might be required is not optimal. Even if the OIC conducts the assessment quickly and then deploys with the other firefighter, the responsibility for effective coordination of the incident is compromised until additional resources arrive. It has been identified through this research, that the organisational culture of many fire services permits (or accepts) OICs to be involved in front-line activities. This, of course, raises question as to the integrity of the overall incident command and control, because if the incident commander is occupied with other duties then there is effectively no-one evaluating the progress of the incident, operational effectiveness and dynamic risks that might arise, as well as incident critical information. In such cases, general oversight of the incident and the ability to fulfil the command and control tasks that are required cannot be performed effectively.

As emergency incidents are of a dynamic nature, the appointment of a person (or persons) to assess associated risks and incident demands is essential. However, to do this, it stands within reason that Incident Commanders should be in a position to undertake their tasks effectively. This means that they should not be directly involved in hands-on aspects of the incident management process i.e. they should not be firefighting or operating tools and equipment. As OICs are accountable for the health and safety of their staff through the delegation of duties that they assign to them, their decision making process and their ability to assess dynamic risks effectively can only be ensured if they have a holistic view, primarily of the function that they are responsible for, whether that be an resource proportionate incident they are managing in its entirety or a sector of a larger scale incident. The joint responsibility for effective command and control along with staff accountability can be directly affected where staffing levels are lower. If OICs are required to engage in operational activities as a result of low staffing numbers then command and control will unequivocally be compromised, increasing the risk for firefighter safety and potentially hindering the likelihood of a positive outcome.

The Importance of Dynamic Risk Assessment

Certain types of incidents that are frequently attended by fire services are commonly referred to as 'routine incidents. However, it has been argued, especially by senior fire officers, that 'there is nothing routine' about any one incident (Naum, 2012; Shaw and Mitchell, 2013; Kennedy, 2015), and due to their dynamic nature, unforeseeable risks might arise if a thorough risk assessment is overlooked in the initial stages. In recognition of this, there is an accepted code of conduct adopted by many fire services worldwide to assist in their risk decision making process. This code is based on a well-known risk management plan first published in 1987 in NFPA 1500, and later adapted by various services from around the world to better fit their policy arrangements. It states:

- We will risk our lives a lot, in a calculated manner, to save SAVABLE LIVES.
- We will risk our lives a LITTLE, in a calculated manner, to save SAVABLE property.
- We WILL NOT risk our lives at all for a building or lives that are already lost.

Past-experience has demonstrated the risks of an uncalculated intervention at incidents. In an article titled 'Blaina: 22 Years On' that discussed the lessons learned from a house fire in Blaina, South Wales in 1996 where two firefighters tragically lost their lives, the Fire Brigade Union (2018) argued that *'firefighters feel a moral obligation at certain incidents to act immediately where life is threatened, and rescues are required.* They continue to state that, *'it is essential to avoid situations which could motivate or pressurise firefighters to act unsafely in the interests of saving life.'* Indeed, controlling impulsive (instinctive) actions by firefighters was one of the fundamental roles of the Incident Commander identified in LFB's Incident Commander Policy, also identified in Fire Service Manual on Incident Command (2008: p 30). In this case, firefighters, given misleading information of the potential of further trapped victims, re-entered the structure on Zephaniah Way in Blaina, after already performing a rescue of a young boy. Whilst half-way up the stairs to the upper floor they were caught in rapid fire development that trapped them inside the structure, where both perished (Fire Brigade Union investigation Report, 1996). A number of failures were identified as contributing to this tragedy, one of which was the inability of the incident commander to undertake a suitable and sufficient risk assessment of the impending fire conditions due to a lack of training. This point clearly associates the necessity for an appropriate level of training with the ability to make sound tactical decisions that will ensure the balance of crew safety whilst meeting operational objectives.

Another fire in a high-rise building in Southampton in 2010 further affirmed this point. Two firefighters entered the scissor design flat on the 9th floor of Shirley Towers and without using their thermal imagining camera moved passed the fire that was in the living room just above the 9th floor and made their way up to the bedrooms on the 11th floor. Though they had not identified the location of the fire due to the heavy smoke in the property, they proceeded to open windows in the bedroom, thus creating a ventilation path from the fire towards their location, and the open windows. Shortly after opening the windows, the windows in the living room where the fire was located failed. This resulted in the stairway operating as an exhaust vent for the fire that progressed so rapidly, turning the stair enclosure into a chimney, that two out of the four firefighters in the property did not manage to escape and were found within the bedrooms just a couple of feet away from the stair landing and the escape door onto the walkway of the 11th floor. Two firefighters perished in this fire and another two suffered significant burns (Hampshire Fire and Rescue Service, 2013). Through a lessons-learned process, the incident at Shirley Towers identified failures which led to a series of recommendations. Amongst those recommendations, were aspects relating to the risk assessment and decision-making processes, incident command and the re-visiting of Standard Operating Guidelines.

A dynamic-risk assessment, as inferred by the title, is undertaken at an incident as the incident is unfolding. The incident commander must analyse the information as it is received and make a series of informed decisions based on their available resources, the extent of the incident, the risks that it presents whilst also keeping in mind some fundamental priorities such as the protection of life and firefighter safety. It has already been established that in order to make sound decisions, appropriate levels of training are essential. Further to this, as the incident progresses the risk assessments, being dynamic in nature is constantly being evaluated based on new information. Training for incident commanders now incorporates a variation of simulated exercises either in conjunction or independently using classroom based software packages such as XVR and incident ground scenarios (Lamb et al., 2014). The fundamental basis of these training modules is that they provide scenarios where information and incident characteristics can change dynamically. Therefore, training delegates must demonstrate their skills in being able to dynamically assess risks and incident requirements and, by using their organisation's incident command standards, to effectively bring the incident scenarios to fruition by demonstrable command and control (Heldal et al., 2016). The variations in crew capacity of the study sample groups raise question as to how a dynamic assessment of risk can be achieved if the incident commander is directly involved in front line activities either. This process is fundamental to effective fire service operations.

Communications

It is widely recognised that portable hand-held radios form a fundamental part of a firefighter's personal protective equipment and radio communications are now widely seen as instrumental to firefighter safety in the fire and rescue service (Varone, 2003; Killen et al, 2005; Thiel, 2013). However, this has not always been the case and even today it appears to be a more prevalent consideration in countries that adopt a structured approach to identifying lessons that can be learned from previous events where communications have been found to be a contributing factor in a firefighter death, injury or near-miss, and where organisational culture is more broadly focused on firefighter safety.

When writing his article, 'Firefighter Safety and Radio Communications' in 2003, Curtis Varone was a 30-year fire service veteran and the Deputy Chief of Providence Fire Department in Rhode Island, USA. In his article, Varone discusses the gradual transition, from the introduction of portable hand-held radios to the fire service in the late 1960s where they were initially seen as more of a hindrance, towards an understanding around the fundamental role radio communications play in firefighter safety today. Varone states that, in the USA, it was not until relatively recently that radio communications and firefighter safety were linked, with many fire fighter fatality investigations in the 1970s and 1980s overlooking this key contributing factor. He argues that as poor radio communications do not directly kill firefighters most investigations concentrated on the more direct and visible causes like rapid fire progression leading to firefighter entrapment, building component failures and collapses, firefighter disorientation and so on.

The earliest documented case of a radio communications failure contributing to a firefighter fatality in the United States was in Syracuse, New Jersey, in 1978 where four firefighters died in a three-story, timber-frame apartment building when fire erupted out of a void space, trapping them on the third floor (Gagliano et al, 2008: pp 445-446). Approximately 16 minutes into the fire, a weak radio transmission *"Help me"* was recorded on the master fire control tape at the Syracuse Fire Department dispatch office. Approximately one minute later, a second transmission was recorded *"Help, help, help"*. Fire personnel on the scene and in the dispatch office, apparently did not hear these transmissions. However, an observer at the scene with a scanner heard the radio transmission requesting help from the third-floor attic and immediately reported this to a fire officer on-scene. It was not clear what action was taken, but a second alarm[32] was not called for another 16 minutes (33 minutes into the fire), and the first of the fatalities was not discovered until about four minutes after the second alarm was called (37 minutes into the fire) (Varone, 2003).

A further series of similar events are described in Varone's article but also in Thiel's Special Report 'Improving Firefighter Communication' on behalf of the Federal Emergency Management Agency in 2013. It was identified that in every one of these events (amongst many others), a failure in communications served as a primary contributing factor leading to the firefighter fatalities.

---

[32] A second alarm is a stage at which the initial response to an incident is up scaled with further resources. In many countries, the initial response as well as subsequent upscaling is standardised.

- On 1st July 1988, a fire at the Ford Company in Hackensack, New Jersey, claimed the lives of five firefighters when a bowstring truss roof collapsed. David Demers, of Demers Associates, issued a report on the Hackensack fire in 1988, concluding that a major contributing factor resulting in the firefighter's deaths was the lack of effective communications both on the fire-ground and between fire-ground commanders and fire headquarters (Varone, 1996: p 80).

- On September 9, 1989, fire claimed the life of a Seattle fire lieutenant at the Blackstock Lumber Company. After becoming disorientated in the fire, numerous attempts were made to communicate for help over his personal hand-held radio. However, neither the Incident Commander, other personnel on the scene, nor dispatch personnel heard any of these requests for help. Members of the public with private scanners, however, reported hearing these transmissions.

- A church fire in Memphis, Tennessee, on December 26, 1992, killed two firefighters after the wood-truss roof collapsed. Crews at the scene were operating on a fire-ground radio channel that was not being monitored by dispatch personnel or the Incident Commander. The United States Fire Administration Technical Report USFA-TR-069/December 1992 identified that communications were a problem contributing to the lack of organisation at the scene since the Incident Commander was unable to communicate with the internal crews on the tactical radio channel.

Other noteworthy events referenced in relevant literature are:

- The Indianapolis Athletic Club fire on 5th February 1992, (Thiel, 2013: p 5), and;
- The Regis Tower Fire in Memphis on 11th April 1994, (Chubb and Caldwell, 1995).

In the UK, a fire at London King's Cross Underground Station on 18th November 1987 claimed the lives of 31 people, including Station Officer Colin Townsley from Soho Fire Station[33], and injured a further 100 people. This was perhaps the first major incident in the UK where radio communications (and the lack thereof) were identified as a key factor contributing to the difficulties experienced around the management of this fire. In the aftermath of this incident, Desmond Fennell (OBE, QC), was appointed by the Secretary of State in November 1987 to hold a formal investigation into the circumstances of the King's Cross Underground fire. In his report published in October 1988, some of the more prominent issues relevant to radio communications were that:

> *"Staff at stations on the London Underground has not been provided with radio equipment because current portable radios will only operate below ground if there is a continuous aerial system throughout the station. The only means of communication for staff at King's Cross on 18 November 1987 was the telephone or word of mouth. Members of the British Transport Police*

[33] Colin Townsley was in charge of the first pump ladder to arrive at the scene and was down in the ticket hall at the time that the flashover occurred. His body was found beside that of a badly burnt passenger at the base of the exit steps to Pancras Road. It is believed that Townsley spotted the passenger in difficulty and stopped to help her before becoming overcome by smoke and fumes himself (Fennel, 1988 pp: 78-79).

> *and the London Fire Brigade at the scene had their own personal radios, but they did not work between the surface and underground. Officers below ground within the station could not communicate by radio either unless within line of sight"* (Fennell, 1988: p 137).

It was also identified that other than technical issues that were faced by responding crews regarding signal coverage and interoperability of equipment, a number of firefighters did not take radios with them when descending into the underground network. Therefore, there was not a clear understanding of the conditions below ground by crews that were above ground and entering into the station to tackle the fire.

Another more recent event in the UK where two firefighters died was at Shirley Towers, a high-rise block of flats in Southampton on 6th April 2010. Overall communications were scrutinised as part of the investigation process with specific points made regarding the relay of critical incident information received by the control room to attending crews in relation to the location of the fire – which was bypassed by crews entering the structure. If this information had been passed on early, the attending crews would have known where to go to directly locate and extinguish the fire instead of passing it to ascend to the upper floors of the flat. However, it must be pointed out that there are other contributing factors that led to a failure by operational crews to identify the location of the fire early on. Further comments were also made in relation to radio discipline and the importance of its use in safeguarding firefighters (Hampshire Fire and Rescue Service, 2010).

In recent statements made in relation to the Grenfell Tower Fire Inquiry, the UK Fire Brigade Union identifies that communication failures have been a recurring theme in firefighter fatalities throughout the UK. Particular reference is made to the Bethnal Green Fire in 2004 where two firefighters lost their lives. Billy Faust and Adam Meere were the 5th relief crew[34] to enter the basement of a clothes shop fire in Bethnal Green on 20th July 2004 but were without radio communications. A number of uncoordinated actions (mainly in respect of ventilation) led to rapid fire progression trapping the firefighters in the basement. It is suggested that personal radios were issued to every firefighter following this incident (FBU, 2008). General communication failures were also identified on 7th July 2005 London bombings and the Lakanal House fire in 2009 (FBU, 2018). Radio communications were also described as problematic at the Grenfell Tower Fire in London 2017 (Lane, 2018). These incidents, and many others, that have been used in research to emphasize the importance of radio communications in firefighter safety, have identified a series of causal / contributing factors. Thiel's Special Report 'Improving Firefighter Communication' on behalf of the Federal Emergency Management Agency in 2013 incorporates these contributing factors into categories and sub-categories as follows:

[34] In the UK, breathing apparatus crews work in a two-in two-out configuration. BA wearers only stay under air for a safe duration of time calculated by the Breathing Apparatus Entry Control Officer, BAECO (this is now done via telemetry but monitored by the BAECO) and are then replaced by a relief crew. Relief crews will be in teams of two and will rotate as many times as is necessary to ensure the incident is brought under control will interior firefighting tactics.

Technical Issues	Human Factors
Unsuitable Equipment	Radio Discipline
Equipment Failure	Situation Reporting
Inadequate System Capacity	Incident Management
Reference Standards	Cultural issues
Radio Interference	Chain of Command
	Problem Reporting
	Completing the Loop – Two-way Communications

Table 4 – Categories and Sub-categories of Radio Communications Failures

Despite this extensive research, it appears from this study that the value of radio communications is still not fully understood with many fire services not still issuing portable hand-held radios as standard practice to every firefighter. It is understandable where teams of two are working together that at least one is issued with a radio though if something were to happen to that firefighter, then the firefighter without a radio would find themselves in a predicament without the ability to call for assistance. In other fire services, such as in Greece for example, the research data identified that many firefighters, understanding the need for personal safety, purchase their own handheld radios. However, the quality of these devices is at best questionable and there is no guarantee that they are fit for purpose. Most fire service radios are for professional use and very expensive. Prices for professional radios are approximately between 60% - 70% of the average firefighters salary in Greece.

Fire services that have standardised equipment policies would strictly forbid the acquisition of personal equipment that is not standard issue and for which the fire service cannot directly account for. In Greece, such processes, as with other personal equipment purchased by firefighters, is generally overlooked.

Furter insight into the status of radio-communications and other self-purchased equipment is discussed in the data analysis section of This book which is pertinent to the sample fire services evaluated as part the study. Broadly speaking, it is purported that a baseline standard is that every firefighter should be provided with a hand-held radio, and this primarily a matter of personal safety.

Firefighter Health, Safety and Accountability

Firefighter health and safety is a key factor in the resilience and sustainability of a fire service. Without ensuring adequate health and safety standards, firefighters are likely to be exposed to unnecessary risks in the short, medium and long term. Being a firefighter is an inherently risky occupation. The conditions to which firefighters are often exposed put them under significant physical and psychological stress that cannot be managed effectively without forethought for their welfare through a robust health and safety management system. Such a system must have strong roots in the organisation's culture and be reflected through policy and practice, representing one of their core values.

The effects on fire-firefighters health through exposure to harmful substances is an area of much scientific interest. There is research that has found that cancer is higher amongst firefighters than the populations they serve, despite their increased levels of physical fitness (Stec et al., 2018). A study by Daniels et al., (2013) into mortality patterns and cancer incidence in a pooled cohort of 29,993 US career firefighters employed since 1950 and followed through to 2009, showed clear associations with digestive and respiratory cancers, alongside excess mesotheliomas – especially in older firefighters - associated with asbestos exposure. Exposure to harmful substances is part in parcel with a firefighter's occupation, though the risk and extent of exposure is something that has changed over time.

A study by Guidotti and Clough (1992: p 158) into the 'Occupational Health Concerns of Firefighting' found that changes in technology over several decades has meant that the risks and their outcomes will vary in accordance with the era that the firefighters entered the workforce. It is further argued that firefighters who entered the fire service before the 1960s were exposed to smoke of less acute toxicity but lacked personal protective equipment of acceptable effectiveness in comparison with modern standards.

However, a significant contributing factor was that health and safety was not seriously considered by fire services until about 30 years ago, with many firefighters operating in fires and other incidents without wearing their personal protective equipment despite being provided with it (Grimwood, 2008: pp 2-3; Thompson, 2017: p 92). This is still very much the case in some fire services around the world whose culture towards health and safety is poor, complacent, or where there is a lack of understanding of the inherent risks associated to fire service operations. The availability of adequate funds is also a key factor in many of these fire services that discount on safety standards. Furthermore, it is unsurprising that relevant research has found that mesothelioma and other asbestos related conditions are more prominent in older firefighters. This can be linked to relatively recent bans on the use of asbestos and the latency (20 - 40 years) for related conditions to become symptomatic (Frost, 2013).

Conversely, those entering the fire service within the last two decades have primarily been exposed to smoke of greater toxicity, due to the chemical compositions of modern materials, but have had more effective respiratory protection available. It is argued that firefighters who joined the fire service in the last few years may have the benefit of both fire-retardant materials produced under more stringent safety codes and also respiratory protection meeting contemporary standards of effectiveness (Guidotti & Clough, 1992: p - 159). However, despite higher standards of safety, the study by Stec et al., (2018) into Polycyclic Aromatic Hydrocarbon (PAH) carcinogens and the risk they present to firefighters tells a different story. In this research of two UK fire stations, samples were taken from firefighter's personal protective equipment, the fire station areas, the fire engines and other equipment firefighters use during firefighting operations.

Airborne PAHs were also sampled and trapped on sorption tubes from the general office and PPE storage area of each of the stations. The results found that, based on a cancer risk factor of 1 in 100,000, the highest risk of cancer is obtained mainly from the dermal exposure with the neck being the area of greatest contamination. Up to 350 firefighters can develop cancer from PAHs concentrations identified on their clothing. Up to 230 firefighters may develop cancer from PAHs found on the fire engines bodyguard and console[35] respectively. A later study by the National Institute of Standards and Technology (NIST) in 2023 showed that the textiles used in protective clothing worn by firefighters often contain per- and polyfluoroalkyl substances, or PFAS, a class of chemicals that has been linked to an increased risk of cancer and other health effects. PFAS are commonly used to provide water proofing to firefighters personal protective equipment and the NIST research found that higher concentrations of PFAS and greater risk of exposure to firefighters associated to general wear and tear of the protective equipment. PFAS can also be found in firefighting foams serve as surfactants that spread the foam over the burning surface material to cool and

---

[35] This is referring to the Breathing Apparatus type and the associated telemetric breathing apparatus control board used to monitor firefighters wearing breathing apparatus at an incident.

suppress the fire (Clean Water Action, 2020). PFAS containing firefighting foams not only present health risks to firefighters but have also been found to present health risks to the environment, the general population and wildlife as a result of foam run off into drains, water courses and the soil. These chemicals are stronger than any chemical bond found in nature and do not break down in the environment or even in the bodies of people and animals (Roman, 2022).

The U.S Fire Administrator's Summit on Fire Prevention and Control in their workgroup summary in 2024 identified that firefighters have a 9% higher risk of developing cancer and a 14% higher risk of dying from cancer compared to the general public due to PFAS exposure. Approximately 70% of fatalities honoured by International Association of Fire Fighters in recent years have been occupationally related cancer deaths.

Health and safety must be an organisational priority and it should be readily promoted at the recruitment stage. The prerequisites for the suitable selection of firefighters should be based on an assessment, against specific criteria, to determine whether candidates are fit and healthy to undertake the demanding role of a firefighter. Health and Safety can either be seen as one collective element supported by a number of values, processes and procedures, or it can be seen in isolation of one another, where the 'health' aspect will refer to the physical health of the individual firefighter and the requirements for maintaining adequate levels of health in order to remain capable of performing their role effectively. This element of health can be translated into a measurement of physical fitness that is performance based and achieved through exercise, sport or other task related physical activities.

Many fire services around the world set physical aptitude tests at regular intervals to ensure that their firefighters are maintaining the required levels of fitness (Oszkar & Ferenc,

2008). Health is also associated to ailments or conditions that might adversely affect the firefighter in the course of their duties. These might be acquired throughout their career, occur at an incident or might even be hereditary. NFPA statistical research has shown that the primary cause of firefighter fatalities is heart-attacks, with sudden cardiac arrest consistently accounting for the largest proportion of on-duty firefighter deaths in the United States since 1977 (Fahy, Le Blanc & Molis, 2018: p 4).

In Europe, the implementation of health and safety measures is a legal responsibility. The stipulations of this legislation are covered in European Framework Directive 1989/391/EEC which establishes general principles for managing safety and health, such as the responsibilities of the employer, the rights and duties of workers, the implementation of risk assessments to continuously improve safety processes, and workplace health and safety representation. This framework is broadly incorporated into most European fire services and which is covered under internal health and safety policies. However, the implementation of the EEC directive into national legislation was not readily adopted by all EU member states and in some cases where this was adopted, it was not enforced appropriately. In Greece, for example, firefighter health and safety standards were not officially considered until 2003 (Halaris, 2008). Many fire services today invest in ensuring the safety of their staff. They have a PPE decontamination policy which is some cases is even applied directly at incident level. These uniforms then go through a robust decontamination process before being returned to active duty (Fernandez et al., 2023). On-scene safety is also practiced by some fire services that encourage firefighters to be washed down before removing their breathing apparatus sets, wear nytril gloves under their firefighting gloves and P2 face masks[36] so as to avoid contact and inhalation with toxic particles and other carcinogens.

NFPA 1851 (2014) 'Standard on Selection, Care, and Maintenance of Protective Ensembles for Structural Fire Fighting and Proximity Fire Fighting' and NFPA 1581 (2015) 'Standard on Fire Department Infection Control Program' provide minimum requirements for firefighters in the United States for the care and maintenance of turnout gear, including decontamination as well as the requirement to identify and limit the potential of an infectious exposure to fire department members. However, it has been identified that These methods are not universally applied.

Safety can be seen as something that is directly influenced by one's acts and omissions, and the assurance of health is highly dependent on the individual's ability to act in a safe manner which shall be conducive to minimising their exposure to adverse risks. The negative results of this process become apparent, and are usually identified, through investigations into firefighter fatalities or where serious injuries have occurred. However, it is argued, that the investigation of human errors in such cases is often avoided. This is either because they are difficult to quantify, because of cultural issues such as camaraderie or sense of brotherhood (where disclosing errors would be taken as a sign of betrayal), or maybe just out of plain fear of what might be discovered (Kerwood, 2009). Furthermore, it might be the case that some fire services do not have appropriate systems or processes in place for recording accidents and near-miss events to draw valuable or analysable data.

---

[36] These are disposable or half face masks that protect against airborne particles present in the form of powders and dust as well as aqueous mists. They come in three ratings P1, P2 and P3 with the latter providing a higher level of protection against contaminants.

It is generally found that reporting and recording of accidents and near-miss incidents in fire services is problematic and thus an understanding of the actual risks presented to firefighters is indeed limited (Taylor, 2015). This fact was pivotal in the establishment of a campaign and training initiative run by Brussels firefighter, Corporal Pieter Maes named the 'Firefighter Mayday Programme'. This programme aims at introducing a culture around firefighter safety. The premise was that though fire services are generally quite good at training for the undertaking of civilian rescues, there is often little focus on tactics and methods that might be required when firefighters get into trouble. Another project supported by the Institution of Fire Engineers (IFE) in the UK and led by Avon Fire and Rescue Service Officer, Adam Course, introduced the new Firefighter Safety Database that was launched in 2018. The IFE developed this database alongside the National Operational Learning (NOL) arm of the UK's National Operational Guidance (NOG) with the aim of providing information, gathered from around the world, on incidents where firefighters died or where their safety was seriously compromised.

Fire Officer, Major Karel Lambert has also incorporated several firefighter fatality or injury case studies on his website www.cfbt-be.com with the prospect of deriving lessons-learned from these events. The objective of sharing this information is the hope that it will contribute to greatly reducing firefighter casualty rates at future incidents. Other campaigns around firefighter safety generally promote the concept that 'telling storeys saves lives' and that recording, and reporting near-miss incidents are critical to firefighter safety in future events. It is asserted that for a firefighter to survive the danger of firefighting, he or she must know how other firefighters have died (Hoff & Kolomay: 2003: p 1). This was indeed the purpose of the National Firefighter Near Miss Reporting System (NFFNMRS) that was established in the United States in 2005. An article written by firefighter Brad Van Ert in the Journal of

Emergency Medical Services (2008) states that *"the three goals of the Near-Miss Reporting System are to learn from real-life experiences, help formulate strategies to reduce the frequency of responder injuries and fatalities and enhance the safety culture of fire and emergency services".*

In the United States alone, over 80,000 firefighters are injured and around 100 die in the line of duty every year (International Association of Fire Chiefs, 2009: p 5). In a paper written by Lauren Johnson of the Dallas Fire and Rescue Department (2012: p 3) 'Collecting Injury and Near-Miss data can Improve Safety', it states that the failure to collect injury or near-miss data for analysis or reporting means that DFRD are unable to identify corrective actions to reduce firefighter injury and death. Indeed, the recording of accidents and near misses is vital if one is to understand the underlying causes of such events to be in a position to effectively overcome them. A near-miss incident is described by the Occupational Safety and Health Administration (2014) as an event that did not result in injury, illness or damage but had the potential to do so. This information is essential in mitigating near-miss incidents transpiring into actual events in the future. It is argued that if the fire service and safety researchers have comprehensive, consistent and accurate information about who is injured, how they are being injured and what types of injuries occur, then these injuries can be prevented. A more recent study titled 'Near-miss narratives from the fire service: a Bayesian analysis' by the Department of Environmental and Occupational Health (2014) has attempted to code incident and near-miss data recorded through the NFFNMRS in order to provide a higher level of reliability and accuracy to the data (up to 90% accuracy) to enable better group an analyses capability. This study concluded that the application of Bayesian auto-coding methods can successfully code both near misses, and injuries in longer-than-average narratives with non-specific prompts regarding injury.

In true fact, the extent of firefighter health and safety is not yet fully understood on a global scale. Though many fire services have started to record critical accident and incident data in order to be in a position to map the implications of health risks on firefighters against a number of variables such as age, years of service, geographical and incident specific criteria etc. many still do not have any such systems in place. Thus, some fire services have a better understanding than others whilst some might not acknowledge the value in understanding these data in the first instance. It might be the case that some fire services simply accept that being a firefighter is a risky occupation and, hence, do not focus greatly on addressing those risks through safe systems of work, risk assessments, the creation and implementation of SOPs and so on.

Standard Operating Procedures, apart from providing a tactical process map to ensure efficiency and effectiveness for firefighters operating at emergency incidents, also incorporate safe systems of work into the process. Where these might fail, leading to an injury or a fatality, the procedure can be reviewed, and errors can more easily be identified against a step-by-step analysis. The investigation of errors becomes simpler where there is a guide to work against that can broadly determine what should have happened and what did happen, and the outcomes of any discrepancies. In scenarios where there are no standard operating procedures in place, the possibility (and likelihood) of errors is without boundaries. If there are no guidelines to inform operational practices, then freelancing takes precedent and health and safety risk increases exponentially. Despite efforts being made by many fire services around the world to tackle this issue and acquire valuable data on the health and safety risks presented to firefighters, there is still not a consensus on how this is done universally.

Much like with other aspects of fire service functions this research has found that there is a large variation in how fire services operate on a global scale, despite many 'best practice' examples. Moreover, the disparity around matters relating to the chronic health implications on firefighters such as cancer, heart conditions, respiratory illnesses, musculoskeletal conditions (such as damaged knees), strokes, neurological damage and many others is concerning. However, despite evidence of relevant research, it was not until recently that certain health conditions such as cancer were recognised as work-related illnesses.

In the United States, the online portal 'First Responder Centre' has published an article titled 'Presumptive Legislation for Firefighter Cancer'. The first heading is in reference to the Occupational Cancer Legislation which states that,

> *"Generally, in the case of work-related illness or injury, the burden is placed on the worker to prove their ailment is a result of occupational exposures. With the advent of presumptive legislation, that burden shifts; the employer must prove that the firefighter's working conditions were not a significant contributing factor to the development of cancer. With presumptive legislation, the line-of-duty claim, and subsequent benefits, can be automatically approved as long as the specific criteria are met under the state's regulations. A majority of states now cover firefighters for one or more cancers under workers' compensation because of presumptive legislation. In many states, the presumptive legislation contains broad or nonspecific language that can be interpreted to cover any cancer experienced by a firefighter. In other states, only specific cancers are covered. Most commonly those are leukaemia, non-Hodgkin lymphoma, brain cancer, bladder cancer, and gastrointestinal cancer. In states where*

> *presumptive legislation has been implemented, records of occupational exposure have played an often-critical role in ensuring a diagnosis of cancer is covered by workers' compensation. This underscores the importance of agency-wide, comprehensive exposure tracking systems. At a minimum, firefighters can, and should, maintain a record of their own individual* exposures".

Some countries still do not formally recognise cancer as a work-related illness affecting firefighters (Brinchmann et al., 2022) and this is not just restricted to poorer countries.

The issue pertaining to firefighter health and safety is multi-faceted and each element is fundamental to the integrity of fire service organisations. Ensuring adequate health and safety standards is not only a legal obligation but also an ethical one. Although it is accepted that firefighters undertake a dangerous occupation, this should not serve as a scapegoat for relaxed health and safety standards. Academic research has contributed notably to bringing critical health risks for firefighters to the fore, but there needs to be buy-in at senior management level to ensure that unnecessary health and safety risks are avoided. This applies to the short, medium and long term health and safety issues currently faced by fire service organisations. However, the data drawn from the sample groups is not encouraging, at least in the case of south European fire services.

Chapter 4 – The Mantzios Fire Case Study

Introduction

Greece suffered significantly as a result of the economic recession. Firefighters (and other emergency workers) took pay cuts whilst at the same time taxation increased markedly. Nonetheless, the fire service, in 2012, recruited approximately 4000 contract firefighters on 5-year contracts who, despite being employed for forest-firefighting and forest fire protection, were being utilised as general duty firefighters. Most of them did not have suitable equipment for the incidents they were attending and they were not appropriately trained.

However, they came to fill a significant resource gap, and so it appeared that the government was content that the numbers of firefighters provided to protect the citizens of Greece was sufficient. Of course, this was not so simple, and this trade-off did not come without expense. In reality, morale was extremely low, busy metropolitan fire stations were left severely understaffed and full-time firefighters felt insecure as their contract counterparts were not adequately prepared, leaving them feeling exposed at incidents. Staff was over-worked; especially during the summer season with many firefighters having accumulated over 300 days of untaken leave. The support from volunteers was declining as work, family and making ends meet became a priority with minimum wage dropping and unemployment increasing[37].

In 2014, the Hellenic Fire Service was overwhelmed by limitations. These limitations were in reference to the extent and quality of training, operational capabilities mainly due to a

[37] Over 20% in February 2018 - http://www.in.gr/2018/04/04/economy/oikonomikes-eidiseis/eurostat-protathlitria-stin-anergia-ellada/

lack of formalised processes, Standard Operating Guidelines and an Incident Command system, the quality and number of staff and absence of a reliable radio communications systems etc. However, it appears that these limitations were not fully acknowledged, perhaps to a lesser extent by the firefighters themselves who had to work with what they had, but apparently, greatly so by the government officials.

This next section will analyse a firefighter fatality that occurred at a building fire in Athens, Greece, in September 2014. This case study has been chosen due to the significant number of results with a negative weighting by Greek firefighter respondents when compared to all other sample groups in this study. The analysis of this case study will identify key points of failure at this incident but more broadly in respect of organisational culture, making reference to fundamental points identified in the body of this thesis. The case-study analysis will demonstrate the catastrophic results that can be experienced by a fire service where there is a lack of formalised processes and procedures to inform fire service operations. It will be asserted and, indeed, demonstrated that when such failures occur, they are often systemic in nature rather than as a result of individual errors. This is certainly the case in the forthcoming incident and in respect of the Hellenic Fire Service as an organisation in general.

It must be noted, that due to a reluctance of firefighters to formally present evidence is support of these findings, reliance has been placed on a series of informal interviews with firefighters and officers. These interviews have taken place as part of two separate studies in 2009 and 2023 by Worldwide Fund for Nature (Greek Division) the former being pertinent to the volunteer firefighting system in Greece and its role in the Civil Protection mechanism[38] and the latter a lessons-learned

---

[38] Pantelides, K. and Tziritis, E. (2009). *Volunteers, Forest Fire Protection and Forest Firefighting: A Proposal for the Effective Inclusion of Volunteers in Forest*

analysis of the catastrophic Evros - Alexandroupolis Fires in 2023[39]. The author was one of the principal authors and investigators in both of these studies and data has been drawn from these to support the narrative of this chapter. Furthermore, the arguments will be further supported (although not explicitly) in the data drawn from the Hellenic firefighter respondents. Lastly, important parts of the narrative in this chapter are derived from the author's 25 years' experience as a volunteer firefighter in the Hellenic Fire Service. In this regard, a level of bias must be factored although the arguments put forward are generally accepted as accurate and this has been determined and cross-referenced through the aforementioned studies as well as from the data drawn from this research. Notwithstanding this point, a subjective bias cannot be precluded although specific points are evidenced where possible.

Protection. Athens: WWF Greece.

[39] Tziritis et al., (2024). Learning from the Evros Fire: A review of the forest firefighting mechanism and recommendations for improvement. Athens: WWF Greece

Fatal Fire Incident

On 12th September 2014 at 20:59:24, the Hellenic Fire Service emergency control room received a call regarding the outbreak of a fire in a ground-floor shop, containing plumbing materials, with flats above it. The property was located at number 4 Ippokratous Street in the area of Palaio Faliro, just a couple of miles south of the centre of Athens.

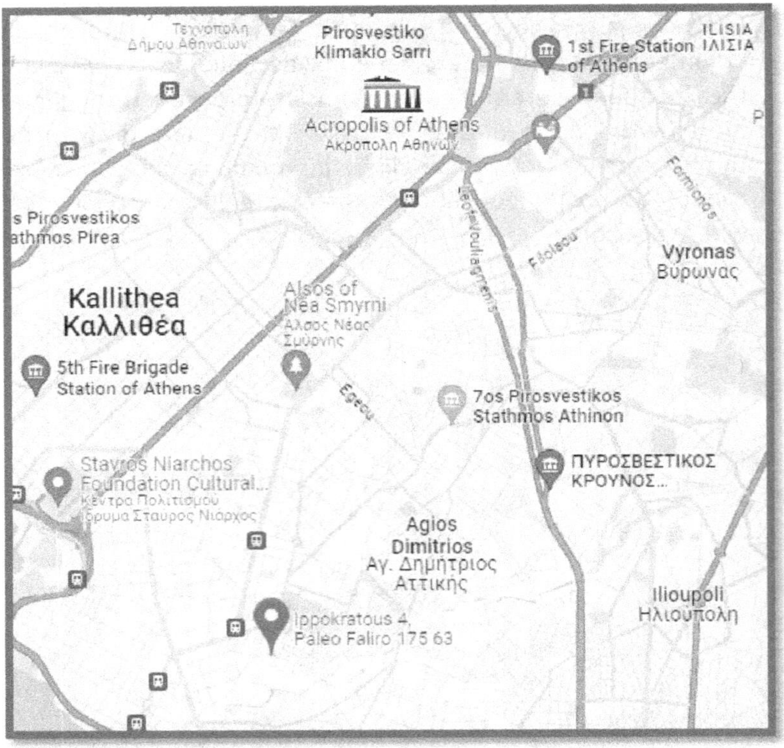

Image 1 – Location of Incident and nearest fire stations

At 21:03:02, the 5th fire station of Athens was mobilised to this incident and responded with two fire engines, as per regulations for fires. The crews comprised four personnel in the first engine (a shift officer, a contract firefighter, a trainee firefighter and a driver); and two personnel in the second engine (a firefighter, and a driver who was also a trainee). It is alleged that this incident was, in fact, under the operational jurisdiction of the 7th fire station of Athens and not of the 5th.

Between 21:02:26 and 21:10:08, the Fire Service control room received a flurry of calls from members of the public reporting that people were trapped in the building; one of the calls stated that a woman has been rescued from an upper floor and was suffering from smoke inhalation. The general picture conveyed by the public was that there was an element of urgency related to this incident. Based on this information, the control room advised the responding units at 21:11:20 via radio to search for trapped victims upon their arrival but did not, however, upscale the resources to the incident. The control room then proceeded to contact the National Ambulance Service at 21:12:58 to request the mobilisation of a unit to the fire to treat a victim suffering from smoke inhalation.

The building was of solid concrete construction with a shop on the ground floor of approximately 100m^2. There were two storeys above ground level, with a main single stair to the left side of the building serving the flats (one on each floor). The main stair did not connect at ground level with the shop. This building was owned and lived in by two generations of family. The oldest of the family, KT, owned the building, and his son, GT, managed the plumbing shop on the ground floor. The son, GT, and his wife, DS, lived in the second-floor flat, and the residents of the first-floor flat were not present at the time of the fire. Apart from the two flats above the ground floor, there was also a mezzanine level constructed inside the shop, with an open stair connecting to the ground floor shop. The mezzanine floor had been converted into a flat spanning approximately 80m^2 over the shop floor area and was the residence of the building owner, KT, who lived there with his wife, AT (the woman rescued by members of the public in the early stages of the fire, prior to the arrival of the fire service). Parts of these 80m² were also used as storage for the shop downstairs. Other than from the internal stair inside the ground floor shop, the mezzanine flat could be accessed via the main staircase of the building that connected all floors from ground to roof level. The fire appears to have started at the rear of the shop on the ground floor, spreading quickly due to the nature of materials (plastics, PVC, fabrics, paints and compressed gas canisters).

Image 2 – Front face of building and location where Mathaios Mantzios was found

Image source (Eurokinissi - Katomeris, K., 2014)

Upon arriving at the incident at 21:13:49, the shift officer, Mathaios Mantzios, requested an ambulance and, at 21:14:48, asked the control room to send further resources. At 21:15:38 the control room dispatched a water tender from the 7th fire station of Athens - that arrived at the incident at 21:24:07 - and at 21:17:39 it dispatched the on-call incident commander[40] who arrived at the incident at 21:40. At about 21:15, Mantzios and the contract firefighter, with whom he was crewed, made their way into the building via the main staircase to search for DS who was trapped on the second floor of building whilst the remaining 4 firefighters made entry into the shop on the ground floor to tackle the fire. Two attempts were made to access the upper floors of the building as, on the first attempt, it was attested that the stair was heavily smoke logged with zero visibility and the team were unable to find their way beyond the first-floor landing to the second floor. This forced them to retreat outside, but pressures by family and members of the public to rescue DS from the second floor governed their decision to make a second attempt to access the second floor. Mantzios and his colleague made their way up to the second floor and found the door to the flat open. Upon entering the flat, it appears that they separated whilst searching for the trapped victim and met again on the balcony facing Ippokratous Street.

It was stated that they only realised that they were on the balcony due to the sound of the pumps from the fire engines that were in operation, and also because of the balustrade that they could feel with their hands. Otherwise, the smoke emitting from the fire in the shop below was so thick that they had very limited visibility even when outside on the balcony. Whilst the search crew were making their way out of the flat, they made a decision to head upstairs to the roof to check if anyone had sought refuge there. Attempts to access the roof were made extremely difficult by large amounts of storage obstructing their route. All of a sudden, there was a reported massive increase in temperature at their location to the extent that the two firefighters started to burn. They started to hastily make their

[40] In Greece, the Hellenic Fire Service has a 24-hour officer on shift who responds to incidents, where deemed appropriate by Control Room, and who assumes the role of Incident Commander until (and if) a more senior officer attends.

way down the stairs when the contract firefighter fell just below the second-floor landing. It is at this stage that both firefighters became separated.

Smoke, heat and fire had spread from the ground floor via the open stair into the mezzanine floor flat and through the open door discharging into the main staircase, venting upwards towards the roof. The door to the mezzanine flat had not been identified during the team's ascent to the upper floors on either of the two attempts though it was reported in the inquiry that both firefighters commented to each other that they thought it was strange that the stair was so heavily smoke-logged. Because the door, opening onto the roof, was closed, this allowed for the heat and smoke to accumulate at the highest point of the building before starting to fill this space (the stair enclosure) and bank downwards. It was reported that the heat was so intense that the combustible materials in the stair above the second floor, including a small outhouse building on the roof eventually became involved in fire.

There are also reports from external crews regarding three explosions in the shop and the rapid escalation of fire to involve the entire ground and mezzanine floor within seconds that appears to coincide with the precise time that the rapid increase in temperature was experienced by Mantzios and the contract firefighter in the stair enclosure whilst attempting to access the roof. As such, attempts to exit the building via the main stair became increasingly difficult due to the extreme conditions, which had turned the stair enclosure into a chimney. The contract firefighter, whilst lying on the stairs below the second-floor landing tried reaching out onto the landing afraid that Mantzios had somehow lost consciousness, but after 2 or 3 attempts and with the heat rising unbearably, the contract firefighter made his way down and out of the building.

In the meantime, Mantzios had re-entered the second floor flat and made his way to the balcony, facing Ippokratous Street. From 21:27 onwards the following sequence of radio dispatches took place between Mantzios and the control room:

- The first communication was at 21:27:29, requesting a turntable ladder. In the background, it becomes apparent that Mantzios's low-air warning whistle on his breathing apparatus set was sounding.

- 21:28:24 – "Control, I am the shift officer. I am on the balcony and running out of air, dispatch the turntable ladder to rescue me".

- 21:28:54 – "Control! The turntable ladder". His low-air warning whistle on his breathing apparatus could still be heard.

- 21:30:52 – "Control, I am the shift officer, I am on the second floor, lying down on the balcony, I need the turntable ladder urgently. I have run out of air".

- Control then responded stating "We cannot hear you clearly. What exactly do you need? What exactly is going on?"

- Mantzios responds stating "Aerial platforms and the turntable ladder. I'm going to die.....from the balcony....do you copy? I'm going to die."

From 21:29:11, the control room started to upscale resources for this incident, first, by alerting the turntable ladder from the 1st fire station of Athens that arrived at the incident at 21:49:44, and then a series of water tenders, specialist appliances, and the fire service ambulance, bringing the total number of vehicles at the incident to 13, with 40 personnel. At 21:32:41, a mobile patrol unit under the jurisdiction of the 1st fire station of Athens heard that there was some confusion over the specific messages sent by Mantzios, and that the control room had not fully understood the severity of the situation. The patrol crew took initiative and re-iterated the full message over the radio identifying that Mantzios is in an emergency situation. The control room confirmed that they had now received and understood this message.

As a result, from 21:33 onwards, attempts were made from the control room to communicate with senior officers attending the fire, as well as with units at the incident to inform them of the emergency relating to their trapped colleague, in order to instigate action towards his rescue. All senior officers were alerted via telephone from the control room, in order of seniority (lowest to highest), starting with:

- The Fire Chief of the 5th fire station of Athens, who was alerted at 21:31:04 and arrived at the incident at 21:51:57.

- The Fire Chief of the City of Athens, who was alerted at 21:33:07 and heard over the radio at the incident at 22:08:17, although it is anticipated that he might have arrived earlier; and

- The Fire Chief of the District of Attica, who was alerted at 21:40:48 and is believed to have arrived at the incident at around 23:03:02.

Based on the incident investigation report and information drawn from telephone and radio communication recordings, there appear to be some discrepancies with regard to the times of arrival of senior officers at the incident. Furthermore, there also appears to be a failure to assess the severity of this incident, at least during initial communications between the control room officer and the senior officers responding to the fire. Persistent communications were also made by the control room to all staff on the incident ground via radio in relation to efforts made to rescue of Mantzios at:

- 21:41:35,
- 21:44:10,
- 21:46:54,
- 21:59:59,
- 22:05:57 and
- 22:14:24.

These efforts turned out to be unfruitful because it was not until much later that there appeared to be any formal line of communication established between the control room and someone on the fire ground in relation to the implementation of actions taken to rescue Mantzios. This is not to say that they did not occur although there is a lack of clarity in the incident inquiry conducted by the fire service and from the statements made by operational firefighters that were reviewed as part of this case study analysis. In the internal fire service inquiry into the incident, witness statements from the attending crews and officers at the scene indicated that numerous efforts had been made to rescue Mantzios.

However, all of these efforts were focused on gaining access to the upper floors via the central staircase, a strategy that was unfeasible seeing as the stair shaft was serving as an exhaust vent for the fire. It appears that it was not until later in the incident that the door from the mezzanine floor leading into the central stair was identified and measures were taken to control the fire from this location as well. Furthermore, when a senior officer was asked why they had not attempted to extend a ladder to reach Mantzios, he replied that *'we weren't sure whether it would have reached'*. The incident report indicated that the balcony of the second floor was just over 8 metres in

height whereas a number of appliances (fire engines) at the incident were equipped with 9 metre extendable ladders.
At 21:59:59, the turntable ladder contacted the control room via radio stating that they cannot access the building from Ippokratous Street and that they needed the hydraulic platform. This is a smaller vehicle with a long articulated arm and a caged platform which can more easily negotiate overhead obstructions such as power lines, trees, etc. The turntable ladder crew continued, *"We cannot access the building on the main stair side. The heat is too intense."* At 22:28:59, the Chief of the City of Athens was heard over the radio stating, *"we are going for the colleague"* (meaning Mantzios), but at 22:28:59 he was heard again stating, *"control, we are trying to make our way up, we cannot go up."*

At 22:29:13 a telephone recording between officers at the incident and the control room identified one officer saying *"...we can't operate. We can't even get a ladder up."* At 23:19:08 the rescue of Mathaios Matzios was finally completed using the hydraulic platform, but he had sadly succumbed to smoke inhalation and was found lying on the second-floor balcony facing the street where most resources were located just 8 metres below. GT's wife, DS, whom Mantzios was trying to rescue, was also found dead as a result of smoke inhalation, inside the 2nd floor flat under a table. Mathaios Mantzios was 33 years old when he died and had joined the fire service in 2007. DS was 44 years old, and a mother of two young children.

In the aftermath of the incident, an investigation was commissioned but conducted internally by the Hellenic Fire Service. The results of the investigation concluded that Mathaios Mantzios's death came as a result of his attempts to rescue the trapped victim and that the events leading to his entrapment were unknown. There is hardly any mention of any failures (systemic or otherwise) or any other factors that might have contributed to the fatal outcome of this incident. Most importantly, however, there were no 'lessons learned' identified, lessons that any member of staff could reflect upon to ensure that similar errors could be avoided in future operations. Moreover, the significant findings of the investigation did not lead to any effective changes within the fire service itself in relation to safeguarding procedures for firefighters on the incident ground apart from a policy that was created in 2016 around testing and maintenance of breathing apparatus. Needless to say, that the breathing apparatus was not a contributing factor in this incident. Subsequently, it was concluded that there were no accountabilities to be held and the case was archived.

It is not the purpose of this book to place blame on particular individuals, after all, this is the duty of a court of law. It is, however, of great interest to identify and draw out wider cultural issues and systemic failures, which support the context of this study. Most importantly, it will become apparent that where a systematic approach supported by documented and robust procedures are not employed, it becomes increasingly difficult to identify points of failure, lessons learned and subsequent actions to avoid or prevent similar errors from occurring again in the future. The generic issues that can be identified as contributing to the outcome of this incident could be considered as the following:

- The inconsistency of the first responding crews, in terms of staffing numbers, competency and training.
- Significant issues with communications.
- The lack of any documented procedures for: initial risk assessment / scene size-up; Standard Operating Guidelines governing operational engagement and decision making; incident emergency procedures, etc.
- A lack of leadership and structured incident command system.

- A failure to implement a systematic approach to dealing with the incident, and the prioritisation of operational tasks.
- A failure to productively investigate the actual failure mechanisms of this incident in order to ensure similar events could be avoided in the future.

It is worth breaking down the fire service response to this event into thematic areas, so that each contributing factor can be analysed in turn in its broader sense. It must be assumed, that where such failures occur, that these will we be reflective of the organisation across a range of incidents, large or small, that it is called to deal with. As such, though this analysis of events is in relation to a structural fire, similar response issues can be reflected across a range of different incidents highlighting the prominence of organisational culture and legacy practices.

The Hellenic Fire Service and critical factors contributing to the Mantzios Case Study

History of the Hellenic Fire Service

The Hellenic Fire Service (HFS) was officially established in 1930 under the Ministry of the Interior, thus replacing the fragmented municipal and military firefighting services that existed previously (Hellenic Fire Service, 1980). Its formation marked a strategic move toward a centralized and professional national firefighting organisation, designed to unify command structures, improve resource allocation, and enhance operational consistency (Papaioannou, 2021).

In the decades following its establishment, the HFS evolved through significant institutional reforms. Post-World War II reconstruction and the challenges of rapid urbanisation prompted a broader remit for the service, which began incorporating civil protection responsibilities (Hellenic Fire Service, 2023). The devastating forest fires of the 1990s, particularly in the Peloponnese and Attica regions, exposed major deficiencies in coordination, equipment, and training—prompting reforms aimed at improving emergency preparedness and inter-agency collaboration (Henderson et al, 2005).

In 1998, the responsibility for wildfire management was transferred from the forestry service to the HFS under law 2612/1998, dramatically expanding its operational scope. This shift, however, occurred without a proportional increase in funding or training, leading to systemic challenges that persist even today. In recent years, the HFS has faced increasing pressure due to climate-induced fire severity, as illustrated by the 2007 and 2023 wildfires.

Consecutive fire seasons in 2009, 2018 and 2021 were also characterised by large and catastrophic events (Xanthopoulos et al., 2021), although the most extraordinary was the fire in Mati on the East Coast of Attica in 2018 that claimed the lives of 104 civilians within a couple of hours. The ferocity and severity of this fire within such a short space of time was unprecedented in Greece (Efthimiou et al., 2020) although significant failures in the emergency response mechanism were also revealed in the post-fire investigation.

Despite incremental improvements in equipment and cross-agency coordination, critiques persist about the lack of strategic planning, data use, and institutional learning (Organisation for Economic Cooperation and Development, 2024) that ultimately hinder modernisation and development. These criticisms have been further corroborated through several post-fire reports by WWF Greece (2008, 2009, 2018, 2021, 2023) the Mantzios Fire Investigation report by Andrianos Gourbatsis in 2014, the Independent Committee report by Goldhammer et al. (2019)

into the Mati fire, as well as other academic studies in this regard.

The HFS remains a militarised service under the Ministry for Climate Crisis and Civil Protection. Its historical development reflects broader issues in Greek public administration, such as centralisation, political patronage, and limited transparency—factors that continue to influence its quality assurance mechanisms and capacity for reform (Toman, 2024).

Weight of Response to the Incidents

It has already been established earlier that the weight of response to any given incident is of critical importance. Many fire services base this response structure on a variation of factors such as risk, property type and, most importantly, the critical tasks that are likely to be required by the firefighters to effectively bring the incident under control. Reference has also been made to the two most common types of response systems (reactive 'conditions-based response' and pro-active 'pre-assigned task-based' response.)

Other fire services employ a purely experience-based intuitive approach without necessarily qualifying whether their response system is sufficient or not. In an attempt to establish the importance of standardised response, a redacted version of the Mantzios case study was sent to a number of operational fire officers in Belgium, London, Spain, Australia, the Netherlands and Germany, in order to acquire the minimum standard response configurations from their respective fire services that would have been dispatched to this incident. The information that was sent included the time of the incident, details regarding the structure and its layout, the information provided to the control room by members of the public and information provided to the first attending crews upon arrival. It also provided pre-fire and post-fire pictures of the structure and incident from all sides for reference. No other information was provided.

The officers were asked two questions to which they had to respond:

1. *Based on the above information and the pictures provided, could you, to the best of your ability, briefly discuss the minimum resource (appliances and firefighters/officers) that you would deem sufficient in the primary response to this incident and what roles would need to be undertaken once on scene?*

2. *Is your response governed by internal policy and / or Standard Operating Guidelines?*

Karel Lambert - Major for the Brussels Fire Service, with 18 years of experience. He is also a part-time professor at Ghent University for the MSc in Fire Safety Engineering.

"Brussels Fire Service would dispatch 2 engines (6 Firefighters in each), 2 ladder trucks (2 Firefighters in each), a command car (2 Staff), an ambulance (2 Firefighters) and a medical crew (with a doctor) as standard response. The medical crew is sent by the military hospital.

The first engine on scene would establish fire attack. If possible, they would also start search & rescue.

The second engine would help with fire attack, search & rescue. If necessary, they would perform water supply and ventilation".

Steve Dudeney - London Fire Brigade Borough Commander with over 30 years of operational experience (since retired).

"An initial call to this fire would attract the attendance of the two nearest pumping appliances. One of these would have to be a Pump Ladder (pumping appliance with a 13.5m ladder) and a Watch Manager in charge. The second would be either a Pump (same but with a 9m ladder and a Crew Manager in charge) or another Pump Ladder.

Upon receipt of multiple calls (more than three calls) a Station Manager (SM; First 'chief' level) and a Command Unit (CU) would have to be sent. It is down to the senior officer at LFB's control centre (Operations Manager) but it is almost certain that the volume of calls received (no specific number but the paper mentions a "flurry of calls") would mean they would use their discretionary power to send an additional two pumping appliances. This is especially likely as one of the calls had mentioned a woman had been rescued by a member of the public. LFB Control would inform LAS (London Ambulance Service) Control who would send an ambulance and most likely an Officer to the scene.

So, it is likely that two pumping appliances would be despatched immediately with a likely arrival time of 5-6 minutes for central/inner London and two more would be en-route within another minute or so with all four likely to be in attendance by 10 minutes from the call. Four pumps riding 4/5 each would mean 16-20 Firefighters (this includes the driver and officer on each), then a station manager and a command unit".

Juan Carlos Campana Lopez - Sergeant in the Fire Service of Madrid in Spain with 29 years of experience

"In my Fire Service, in a situation like this, with that initial information, and with multiple calls entering in the Control Centre (112), our automatic system would propose the following resources:

- *1 Pump/ Engine: 6 Firefighters + 1 Sergeant + 1 Driver*
- *1 Ladder: 2 Firefighters+ 1 Driver*
- *1 Command Vehicle: 1 Sub-officer + 1 Firefighter + 1 Driver.*

TOTAL in first response: 1 Sub-officer + 1FF + 1 Sergeant + 8 FF (4 teams of 2) + 3 Drivers = 14 staff
Surely, and with multiple calls for the same incident, the Officer in charge of the service that day would upscale these resources with an additional response of:

- *1 Pump / Engine: 6 FF + 1 Sergeant + 1 Driver*
- *1 Ladder: 2 FF + 1 Driver*
- *1 Command Vehicle: 1 Officer (himself) + 1FF + 1 Driver.*

TOTAL in second response (or simultaneous response): 1 Officer + 1 Firefighter + 1 Sergeant + 8 Firefighters (4 teams) + 3 Drivers = 14.

Nowadays, and after so many years of asking for SOPS, today, we have some of these, but not as many as we would like. For us it is very important to have SOPs".

John McDonough - Duty Commander at Inner West C Platoon at Fire & Rescue New South Wales. He is also an international leader in firefighter training and firefighter safety.

"Our standard response to this incident would be a 2nd Alarm. As noted below, this would then be upgraded to a 3rd Alarm by first arriving Officer.

2nd Alarm – 4 x Pumpers for Operations, 1 x pumper for the Rapid Intervention Team, 1 x Aerial Platform (usually a 37m Bronto Skylift) and Duty Commander for incident management. A standard Pumper crew consists of one Station Officer and 4 firefighters. Aerial platforms have 2 operators.*

On arrival, the first arriving Station Officer would assume command and become the Incident Commander (IC)".

*A Rapid Intervention Team is designated as an emergency response facility for firefighters working inside structures or in hazardous atmospheres wearing breathing apparatus. Should an emergency arise, the Rapid Intervention Team will be deployed to perform a rescue of the firefighters in distress. This is generally the sole purpose of this team when at a fire incident and they will not be involved in anything else other than standing-by in anticipation of a firefighter emergency.

Shan Raffel - Station Officer in Queensland Fire and Rescue Service (Ret) with over 34 years' experience in front line firefighting. He is also an international leader in firefighter training and firefighter safety.

I believe it will be easier to start the discussion by answering the second part of your questions first. Is your response governed by internal policy and standard operational procedures / guidelines?

Yes. We have pre-determined response plans. These are grouped into a number of broad incident categories. For example, we have "Structural", "High Risk Structural", "Nursing Homes" etc. A structural response is a minimum of 2 Pumpers. They are crewed with 1 officer and 3 Firefighters. Most of our stations are 1 pump stations and the first appliance will arrive in 6 to 9 minutes and the second appliance 2 – 4 minutes later.

The next level of pre-determined response is for buildings that have been risk assessed as requiring a greater weight of response. The next level of response is what we refer to as Greater Alarm Response System. On receipt of the first emergency call to the communications centre, a "Structural" response consisting of 2 Pumpers would immediately be dispatched. They would also notify the Ambulance Service, Police and the Electrical Supply Authority.

If the communications centre received multiple calls indicating persons were trapped, they would relay this information to the responding crews. In that case it would be highly likely that the communications centre would send a 3rd Pumper. If the Comms centre was informed that the housing levels were above a commercial shop that would also have triggered an upgraded response before the arrival of the first crews. On receiving this sort of information en-route, I think it would be fair to say that most of our Station Officers would have called for an extra Pumper or even called a Second Alarm which would give 4 Pumpers, 1 Command Officer and 1 Control Vehicle.

On being informed en-route that there were persons in need of rescue I would have upgraded to a Second Alarm.

Hans Neiling - Senior Compartment Fire Behaviour Training Instructor at CFBT-NL and a Captain (Ret) at Zuid Limburg Fire Service in the Netherlands with over 30 years of service.

"With a confirmed fire, they will send out 1 pumper with a 6-firefighter crew and 1 ladder appliance with a 2-firefighter crew. An officer (battalion chief) will also be dispatched.

If there are known persons trapped inside the building, the incident will be escalated to a middle alarm that would mean that another pumper with a 6-firefighter crew will be dispatched. If it would be confirmed that flames are exiting the building, then they would mobilise another pumper with a 6-firefighter crew.

- *The response time for the 1st pumper will be 8 min from alarm until on the scene.*
- *The response time for the 2nd pumper will be 12 min until on the scene.*
- *The response time for the 3rd pumper will be 15 min until on the scene.*

The reason for this system is as follows:

- *1 pumper for fighting the fire.*
- *1 for searching for victims.*
- *The 3rd pumper is for assisting both pumpers where needed.*

If there are possible victims involved, it will always be 2 pumpers as a minimum response. This is the standard operating procedure for the Dutch Fire Service".

Tonio Brix - sub-officer with Hamburg Fire and Rescue Service in Germany. He has been serving as a full-time firefighter for 7 years and has an additional 14 years' experience as a volunteer firefighter, serving in the role of sub-officer.

"For a structural fire of this type with persons trapped, the standard response for Hamburg Fire and Rescue Service would be:

- *2 Fire Engines with 5 firefighters and an officer in each (Total of 12 Firefighters)*
- *1 Turntable ladder with 2 firefighters*
- *An incident commander with command vehicle*
- *An ambulance*

If the incident needed to be up-graded, control would dispatch a further two fire engines with 6 personnel in each and an additional turn-table ladder with 2 firefighters. Ambulances would be dispatched based on the requirements of the incident. In the initial response, to this incident, Brix would require 2 fire engines and a turntable ladders with 12 to 14 firefighters in total, 2 officers, one to coordinate search and rescue activities and one to coordinate firefighting, as well as 2 ambulances.

These response procedures are governed by national standards".

It becomes clear from the respondents that their respective fire services not only have clearly defined procedures regarding primary attendance to incidents but also pre-defined minimum staffing levels. It has even been demonstrated that other than considerations for the initial response, the upscaling of further resources constitutes a well thought through process that is based on a number of task critical indicators (CTPIs), time and risk dependent factors. What is interesting is that every country has a different formation to the resources they would dispatch to an incident with the case-study variables without any necessarily being better or worse than the other. The common theme is that each response formation is justified on the above dependent variables. However, it has been acknowledged that not all fire services will have the capability to facilitate defined staffing levels, especially those that rely on retained (on-call) or volunteer firefighters. Many fire services around the world have reduced their minimum staffing levels down to three firefighters per fire engine and even two in some cases.

In the Mantzios case study, it is apparent that the initial response, when compared to the other fire services, only had very limited staffing (4 in the first appliance and 2 in the second appliance) allowing for a total of six firefighters in the initial response. Staffing of such numbers is not uncommon in the Hellenic Fire Service (and assisting volunteer organisations), even at the busier metropolitan fire stations. These figures become even more risk critical when multiple incidents occur and the two responding units, which is the standard response for any fire, road traffic collision etc. need to separate, meaning that, depending on the incident, both appliances are likely to be attending their respective incidents with limited staff. Though it is understood by members of the Hellenic Fire Service that higher staffing levels would allow for greater capabilities and safety of operational crews, this is a challenge that has yet to be resolved effectively and initially came about as a result of the freeze in fire service recruitment from 2001 until 2006.

It is understandable that the lack in human resources was exacerbated over this 6-year period with numerous firefighters retiring, taking valuable experience with them, but without those positions being replaced by a new in-take of firefighters to whom knowledge could be passed down. However, today, it is commonly argued by fire service personnel that this is no longer a matter of organisational understaffing, but as a result of incorrect resourcing of available staff into either non-operational roles or facilitating 'favours'[41] by granting requests for transfer of staff to their hometowns or villages to work at less busy fire stations, resulting in significant understaffing of the busier metropolitan stations.

It is not unheard of, that in some areas of Greece, due to low staff numbers, water tenders or other miscellaneous vehicles will respond to incidents with the driver alone. A two-person crew is commonplace, especially, for patrol vehicles during the forest-fire season (1st May to 31st October), and many of these units will be also utilised to attend other types of incidents (structural fires, vehicle fires, lift entrapments etc.). Forest fires themselves are extremely resource heavy and two-person crews at such incidents can be pushed to their physical limits, putting them at greater risk of injury, exhaustion and so on. This was identified in a recent study undertaken by Worldwide Fund for Nature (Greek division) into the 17 day forest fire in the region of Evros in Greece in August 2023 where low staffing numbers on fire engines was identified as a significant factor in firefighter performance. Only in March 2024 has an internal order been given to fire services to staff all city and forest patrol units with 3 firefighters as a minimum. In terms of resources, in 2015, just a year after the fire where Matthaios Mantzios died, it was noted that there were 115 firefighters per 100,000 population in Greece, 109 in the United States, 80 in France, 48

---

[41] These 'favours' are known as a 'Rouspheti'. This process originates from the Ottoman Empire where Greeks, as the underdog community, in order to acquire any sort of service by the Ottoman Turks who were responsible for public administration, had to pay what was effectively a bribe. This trend became commonplace and formed an essential bargaining technique to ensure preferential treatment against the norm and poor standard of living. It was a method of securing some sort of opportunity, wealth or perceived benefits, and to this date, still forms an integral part of Greek culture especially through the process of any sort of transaction or acquisition for self-benefit.

in Italy, 54 in Germany, 65 in the UK, 37 in Portugal, 31 in Austria, 78 in Poland, 33 in the Netherlands, 52 in Belgium and 92 in Bulgaria (Gourbatsis, 2015: p - 11). Equivalent figures from Eurostat in 2022 show that Greece, Estonia, and Cyprus registered the highest share of firefighters in their national employment, with shares of above 0.4%, while the Netherlands and France had the lowest shares at around 0.1%.

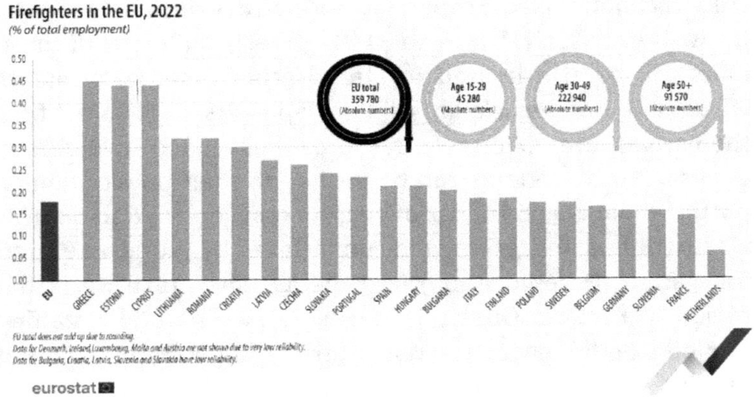

Figure 7 – Firefighter in the EU, 2022

(Eurostat, 2022)

It can therefore be reasonably surmised that resourcing issues in the Hellenic Fire Service has more to do with mismanagement of existing staff. In 2024, there are fire stations in Athens, some of which are among the busiest in Greece, that cannot operate without the support of volunteer firefighters. A significant number of operational staff are engaged in roles where most north European fire services would employ civilian staff such as in administration, call handlers, personal assistants and other office based jobs. Nonetheless, even with 6 firefighters at the case-study incident there is plenty that can be done in a safe and effective manner. However, in order to achieve this, standard operational procedures must be in place to serve as a process map, to identify incident objectives, inform decision making around incident management and the tactical deployment of available resources. The priority is to achieve mission objectives through safe systems of work, and this must be a fundamental priority especially where there is limited staffing or staff of limited capabilities, as was the case in the case study fire.

The Absence of Standard Operating Guidelines

The Hellenic Fire Service, unfortunately, does not have documented Standard Operating Guidelines for the mobilisation of fire appliances, incident management and so on, nor has there been any intelligent risk-based analysis undertaken to consolidate such protocols or guidelines into a structured format. At least, none have been identified as part of this study. There is an internal policy document, Presidential Order 210/1992, that outlines the roles and responsibilities of the fire service organisation with reference to specific roles and functions, but this document is over 30 years old and does not provide the intricate detail required for task specific guidelines. There are several operational protocols currently in place in the Hellenic Fire Service, but these are out of date and not used in practice at operational level. The absence of other documented procedures does not mean that the experienced - based (reactive) approach adopted by the Greek fire service does not work at all. After all, they still continue to save lives and property with relatively low numbers of firefighter fatalities. However, their unstructured, and sometimes unorthodox approach does have significant shortfalls.

There are certain functions, like methods of hose-laying, responses to certain types of fires, or performing miscellaneous functions like lift entrapment rescues, where staff acquire a highly standardised approach to dealing with such incidents due to their frequency and the generic sequence of actions required to resolve them. If someone is trapped in a lift, one simply goes through the motions, and as experience has shown that the level of difficulty will rarely vary, a defined method is identified and used as a panacea for addressing the same problems over and over again. Most actions at such incidents are conducted mechanically and are lacking in formal structure. This is evident from the number of different ways that such incidents are approached by firefighters based on their perception of risk or extent of technical knowledge.

Another fundamental problem is that there is also no process for firefighter accountability in building fires which means that it is unlikely that the location of firefighters is known at any given time at an active incident. This increases the risk markedly as there might be significant delays until it is established that someone is missing or in trouble. The ship *Eleftherios Venizelos* caught fire at sea whilst traveling from Athens to Crete in August 2018. The fire started in the 3rd basement parking level and involved a number of lorries and other vehicles. The ship managed to return to the port of Piraeus in Athens, where firefighters undertook a four-day battle to find and finally extinguish the fire (CNN Greece, 2019). Due to the critical risk that firefighters were placed at, officers created an ad hoc accountability system that involved crews reporting to an entry control officer just outside the ship's garage entrance whose names were documented on a piece of paper before they entered. Upon exiting the ship, they had to again report to the officer so that accountability could be established. This indicates that measures were implemented circumstantially because they were perceived as necessary to ensure the safety of firefighters working inside the burning ship. It is an interesting antithesis as to why such a process would not be considered equally essential at structural fires, such as the one that Mantzios attended and every other incident alike.

In reality, the mobilisation of resources to incidents by the Hellenic Fire Service control room is heavily experienced-based and reactive with very little to work with in terms of documented procedures. Though there are many experienced members of staff working in the control room, it is believed that in the event of a tragedy, such as in the Mantzios case study, their decision-making process cannot be substantiated in a court of law, other than perhaps on grounds of personal judgment. Thus, it is believed that without the support of documented guidelines to which they can demonstrate adherence, it will be very difficult for them to justify their decision-making process on experience alone.

In relation to the Mantzios case, there are some important legal and ethical questions to be considered such as how can an individual control room officer demonstrate, beyond reasonable doubt, that the initial deployment of resources to this incident was, indeed, adequate? Without documented guidelines, how is it possible to identify where in the process errors may have occurred? A key point mentioned in the technical report drafted on behalf of Mantzios's family, was regarding why the 5th fire station was dispatched to this incident instead of the 7th fire station that should have been mobilised instead, as the fire incident was under their jurisdiction (Gourbatsis, 2014: p 15). It was also argued that the crew from the 7th fire station was more competent than the crew on shift at the 5th fire station (ibid). In reality, regardless of which station was sent, both had 6 firefighters available to attend this incident in the initial response and the control room officer would have no method of determining the competency of the crews at either station. In fact, it can be argued that despite the incorrect fire station being mobilised to the fire, the matter of perceived competency should not really play an important role. Fire stations and their firefighters attend fires and, as such, their staff should be adequately trained to deal with any such incident regardless of what station they are responding from. All over the world, it is common practice that fire stations will attend incidents outside their own station ground if the local fire station is occupied at another incident – even in Greece. This comes part in parcel with the resilience targets of fire services.

In the Mantzios case study, making assumptions about one's competency requires a base level from which to compare against, otherwise, as a notion, it becomes intangible. There is no such base level in Greece, at least not in comparison to the formal training systems implemented in north European fire services.

The argument around competency further demonstrates the immense benefits of setting specific criteria and role-specific responsibilities. Similarly, the actions that would be taken to deal with this fire, subject to the different variables at play, the available staffing levels and other risks present, if consolidated into a documented guideline would broadly create homogeneity in the fire service's approach to all such incidents. There are no documented guidelines to establish these principles in Greece and the actions by Mantzios and his crew, it can be argued, were no different than any other crew at any other fire incident attended by the Hellenic Fire Service. With this in mind, it could reasonably be argued that Mantzios's predicament was bought about more as a result of bad timing and bad luck rather than personal errors or flawed decisions. In reality, without the support of such documented guidelines by which points of error can be determined, the weight of responsibility is neither here nor there and people can readily absolve themselves of responsibility on the grounds that they took all necessary action to save Mantzios.

It is not uncommon for such vagueness, disorganisation and lack of accountability to exist throughout all sectors of Greek society. It is, in fact, a deeply rooted cultural issue identified in many studies, affecting all aspects of life including the economy, the country's public health system, the public sector in its entirety and everyday social interactions. Research has identified that there is not a lack of laws in Greece but a problem with their implementation. In his paper, *'Law Making, the Rule of Law, and the Quality of Democracy'*, Danopoulos (2015) found that nearly

50% of laws in Greece are not implemented. A survey by the newspaper 'Kathimerini' (2007), also referenced in Danopoulos paper, further found that though the Greek state possesses an extensive legal framework, popularly elected governments do not have the political will to implement it. As a result, where such impunity exists, the absence of documented processes in some ways, serves a purpose. In informal discussions with members of the Greek fire service around this topic, a common argument that arose numerous times was that 'if there were documented procedures in place, then it would be easier to hold someone legally accountable for their actions or for not following the procedures' and, therefore, something that should be avoided.

Firefighter Competency and Training in the Hellenic Fire Service

The Hellenic Fire Service appears to lack the consolidation of a strong ethos or culture around continual professional development at organisational level. This could be as a result of an inadequate or ill-informed understanding around training needs and a perception of what the fire service wants their firefighters to do. Even today there is no training academy with facilities that are comparable to those in Northern Europe. It is only fairly recently, in 2014, that a mobile training unit was acquired for the purposes of conducting live fire training using propane gas to simulate fire conditions. This training module sits on an articulated lorry and can simulate multiple compartment scenarios for search and rescue and firefighting (Draeger, 2024). Although this training prop is very useful for developing situational awareness skills, familiarisation with working in hot and limited visibility conditions, the overall training capabilities are not directly reflective of real compartment fire conditions. As there is only one of these units currently available (as of 2024), the ability to retrospectively train firefighters throughout the country is limited. It has also been stated by numerous firefighters that due to the restrictions in budgets, this mobile unit is only sent to fire stations that can self-fund its running costs, either through donations, or through money raised by the firefighters themselves.

In March 2024, the fire service approved two more training modules that simulate live fire conditions. These are in the form of modified shipping containers that use solid fuels (either chipboards or oriented strand boards) in a specific configuration to simulate compartment fire conditions with the purpose of training firefighters in managing hot smoke to prevent flashover by creating a safer environment by cooling the smoke and working in zero visibility.

One of the training props is housed at the 6th fire station of Athens and one at the old training academy in Sindos, Thessaloniki. The former was developed by firefighters and officers at the 6th fire station who were trained over two consecutive visits (2019 and 2023) by CFBT-BE that is a private training company comprised of fire officers and fire fighters from Belgium who also train the Belgian fire service amongst many other fire firefighters globally. The training facilities in Sindos were developed by George Bogkias, a volunteer firefighter from Thessaloniki and local firefighters and officers on a voluntary basis. This initiative was the supported with numerous donations in equipment from north European fire services. The first instructors were trained by George Bogkias and Stephane Morizot and were certified by the Yvelines firefighters academy in France (SDIS-78). As of October 2024, a first tranche of trainee fire officers was trained Thessaloniki, followed by Athens in February 2025.

Training manuals and curriculum mostly date back to the 1990s, and it is only thanks to some younger and more enthusiastic instructors that certain levels of innovation, or deviation from standard training, can be implemented. This, once again, is highly dependent on individuals and does not form part of a structured process with a specific curriculum that is targeted around desired competency criteria for firefighters and based on quantifiable evidence. This is further compounded by the recruitment process and the induction of the fire service training academy for both firefighters and officers into the higher education system. This means that the recruitment process is achieved through an examination process that forms part of the national examination system for final year secondary school students wishing to continue into higher education. Following informal discussions with new recruits, it was understood that both the facilities and curriculum are wholly inappropriate, especially for those wishing to become firefighters who are required to attend the training academy for two years. This, since 2023, has increased to four years.

Trainees officers and firefighters are required to complete 4 years in order to achieve equivalence to a university qualification. The problem is that the training material is too generic and not relevant, with many trainees feeling that they have not been given adequate provisions to undertake their job safely and effectively. Many trainees have reported very limited practical training even after their second year in training school. Although anecdotal, it was mentioned that when they complained to management, they were told that they should feel lucky that they have secured a job in the public sector and be grateful for this opportunity. In 2024, it still appears to be the case that hands-on training is very limited. Practical experience and training are supplemented through periodic placements at fire stations but, although they are staffed on front line appliances, they cannot be used effectively at live incidents.

There are, however, numerous good practice examples that can be identified in the Hellenic Fire Service, but it is believed that these are more reflective of personal initiative rather than of organisational culture. Initiatives are taken to progress training, but these are sporadic and do not form part of a greater plan to establish training needs at an organisational level across. Busier and more experienced fire stations, as well as specialist units (e.g., the Special Unit for Disaster Response - EMAK) will have developed effective systems of work demonstrating technical competence for certain specialist areas such as road traffic collision rescue, mountain and rope rescue, rescue diving and so on, but in many cases, these will be identified at shift level.

Essentially, there is often no homogeneity between shifts at different fire stations or even between shifts at the same fire station. This becomes even more complicated when seasonal, volunteer and contract firefighters are introduced into the equation, many of which – especially volunteers - might never become entirely integrated or accepted into the shift dynamics by virtue of their role and title. A paper by WWF-Greece in 2009 that looked at the utilisation of volunteers in civil protection activities identified that fire service culture in Greece is very much orientated around status, with volunteers being regarded at the lowest level in the organisational structure and never fully integrated (WWF 2009). For example, a volunteer firefighter might often be referred to as *'the volunteer'* when on shift at a full-time fire station, rather than by his or her name. Contrarily all full-time firefighters will refer to one another by their name, surname or rank.

On the day of the fire, Mantzios was the longest serving member of his crew, having joined the fire service in 2007. The driver he was crewed with in the first appliance had joined the fire service in 2008. The contract firefighter he was with until his last moments had originally been employed for forest firefighting and had joined the fire service in this role in 2012. The last member on the first appliance was originally a contract firefighter, employed in 2012 for the purposes of forest fighting but had since been employed full-time in August 2014 and was undertaking his practical training as a trainee at the 5th fire station of Athens. In the second appliance, there was one firefighter that had joined the fire service in December 2013, who in his statement reported that this was the first structural fire he had attended in his career. The driver was a trainee with the same status as the trainee riding in the first appliance with Mantzios, having joined the fire service as a contract firefighter designated for forest firefighting duties and subsequently converting to full-time in August 2014 and undertaking his training at the fifth fire station of Athens.

In many fire services around the world, this crew, both in terms of size and competency, would not have been permitted to respond to an incident. Under no circumstances can this crew formation be considered as consistent or appropriate, but it must be acknowledged that this is a systemic issue that is broadly accepted by senior management and reluctantly tolerated by front line staff. This is further complicated by the failure of the Hellenic Fire Service to implement formal and meritocratic competency criteria by which to determine staff capabilities. The result of this is that the extent to which a firefighter is able to perform specific functions will vary significantly and be dependent on a number of subjective factors. For example, experience aside, it would be assumed that any firefighter riding on a fire engine at a busy metropolitan fire station will have the same general breadth of knowledge and similar skill set as any of their counterparts. In Greece, this is not the case.

Who was in Charge of the Incident?

The Hellenic Fire Service does not have any documented procedures for incident commanders and there is, further, no stipulation in its internal policy document, Presidential Order 210/1992, restricting their hands-on intervention at incidents. In fact, in a discussion with a senior Hellenic Fire Service officer (since retired), the very idea of such a possibility was perceived with outrage and it was stated that *'this will never happen in Greece'*. This can have a significant impact on fire service operations, as if the incident commander is directly involved in the incident, then there will be no-one available to manage the incident demands, unless another officer arrives and assumes command. Furthermore, if the incident commander is not physically present, then who will the crew be accountable to and, moreover, who will the incident commander be accountable to if they are actively involved in the incident, and they are also the most senior person on scene?

The incident commander will generally be the most senior officer at an incident or sector of an incident at any given time i.e., with first arriving crews, the officer in charge will usually be the most senior firefighter – in this case Mantzios.

It is broadly accepted that Incident Commanders must:

- Undertake a scene assessment (or size up).
- Identify risks to life and property as well as any other risks that need to be addressed.
- Establish the mission objectives.
- Communicate the mission objectives to the crew when accounting for those risks.
- Monitor progress.
- Adjust tactics where required to satisfy the mission objectives.

The risk assessment process will be considered against a number of dynamic indicators as well as some fundamental principles. One of these principles is the protection of life. This is sometimes restricted to lives of members of the public though protection of life, at least for the incident commander would normally encapsulate the lives of firefighters as well. Examples of some fundamental principles relevant to the safety of firefighters when dealing with structural fires are to (Brandweeracademie, 2019):

- Never commit crews above the fire without the protection of a charged hose-line.
- Never commit crews into a building for search and rescue without the protection of a hose-line or in front of the primary attack line.
- Always isolate or control the fire before placing crews above it.

The latter safeguarding procedures are discussed extensively in the field of fire service development and challenge the traditional school of thought that simply focused on prioritising life rescue objectives without necessarily considering the potential risks that firefighting crews might be exposed to (Lambert, 2012). Risking a firefighter's life to save another was all too often seen as a job-related hazard. Essentially, this revised approach has brought about a better balance in 'risk versus gain' decisions on the fire-ground considering life safety objectives for firefighters as well (Grimwood, 2006). It is no longer considered acceptable to put firefighters at risk to save life and property that are not salvageable, or, generally, put firefighters lives at risk unnecessarily. This is clearly outlined in the principles of NFPA 1500, which have been adopted by many fire services world-wide, as well as other internal fire service policies.

At the Mantzios incident, the objectives that needed to be achieved were broadly established; get water on the fire, perform rescues of persons reported in the building. However, the vital role of the incident commander and their associated tasks would require Mantzios to be outside the building to be able to coordinate the incident effectively. Further to this, the risk presented by the mezzanine door that was open into the main stair was not established early-on so that this critical piece of information could be assessed, and the risk controlled appropriately before advancing above the fire. As Mantzios was actively involved in the incident, there was no-one to relay this important information back to.

In the internal fire service incident inquiry into the Mantzios case, the initial responding crews stated that they felt that there was no co-ordination of this incident from the point that Mantzios had gone missing. Many crew members reported that they were simply following orders mechanically but without knowing who from or even why they were following them as they were unaware of who was in charge. Objectives were felt to be unclear and members that formed part of Mantzios's crew, who were involved in rescue attempts, had evidently become overwhelmed by stress and exhaustion, and coming to terms with the likely outcome of their leader's fate.

This direct involvement of the Incident Commander in fire and rescue operations is commonplace in Greece and forms an integral part of fire service culture. This is often related to a pre-conceived notion that demonstration, or rather validation, of one's competence, bravery, worth and leadership abilities can only be achieved by their direct involvement in the incident itself as 'the first one in', the one leading the way and setting the example for others. Officers who do not engage operationally in a hands-on approach are often referred to as 'ship-wrecks', a term used colloquially to describe someone who is unworthy, incompetent and without courage. The author has many personal experiences throughout his operational involvement as a volunteer firefighter where some of the most senior officers (Incident Commanders) are occupied at an incident whilst undertaking hands-on duties – even at the very end of an incident where damage control functions, like damping down, are taking place.

Effectively, in Greece, competence and leadership, even for officers, is often assessed against the extent of aggression and fearlessness one demonstrates in the undertaking of firefighting and rescue operations. Apart from this important cultural aspect, another reason for the direct involvement of officers in the incident results from both the low levels of staffing and the varying levels of competency – as was the case for Mantzios (as the most experienced firefighter) and his allocated crew. In the fire service report of the Mantzios case study fire, it was identified, in the surviving contractor firefighter's statement, that both he and Mantzios thought that it was strange that the stair was so heavily smoke logged, yet the decision to proceed or withdraw sat entirely with them, and mainly with Mantzios as the most senior firefighter at the incident, as there was no-one else on scene at that time of greater seniority to take control and evaluate that information independently and make informed decisions.

This is where the value of dynamic risk assessment, SOPs, as well as a defined incident command role are most critical. In the Incident Command Policy of LFB, number 408 (2017), it is clearly stated that '*application of the Decision-Making Model to an incident gives the Incident Commander a structured approach to achieving the required outcome through Safe Systems of Work*'. If this incident had occurred in the UK, for example, it is more than likely that such information would have been assessed, evaluated and measures taken to control that risk if it were deemed significant, either by identifying the source of smoke spread into the stair and controlling that before committing crews or even withdrawing crews altogether if that was deemed as the safest option. What is worse is that there was a lost opportunity to mitigate this tragedy from occurring, as two separate attempts were made to enter the property. On their first attempt, Mantzios and his colleague only reached the first floor but decided to retreat due to the smoke conditions and an inability to find the steps leading up to the second floor. Though it is not clarified in the surviving firefighter's statement, it can be surmised that the decision to retreat at this point was driven as much by instinctive reasons as by practical ones. However, pressures imposed on them to rescue the trapped victim by a member of her family governed their decision to re-enter the property which resulted in the tragic outcome. Therefore, much like Blaina in 1996, and as put by the Fire Brigade Union following that event, indeed, it is essential to avoid situations which could motivate or pressurise firefighters to act unsafely in the interests of saving life. In the case of Blaina, information was incorrect as there was no further victim trapped inside the property, and in the Mantzios case, the victim was unlikely to have been alive at the time the fire service arrived on scene – though this can only be speculated. Furthermore, the Incident Commander's role to 'ensure that any tendency for crews to commence instinctive action is held in-check until positive directions are given genuinely makes sense when put into this context.

A concerning reality is that Mantzios actions were not out of context nor were they in contravention of the status quo i.e., what happens at many other fires attended by the Hellenic Fire Service. As it has already been established, officers are more often than not involved in front line activities. It is, thus, assumed that his decision-making process would not have been significantly different than any other officers at the same incident. Of course, due to the dynamic nature of these events it is difficult to determine what the outcome would have been if perhaps there had been a more experienced officer or crew available on that day, or if more resources had been despatched sooner, radio communications had been better and so on.

The importance of dynamic risk assessment should not be overlooked in the Mantzios case study and if one was conducted, in conjunction with having a defined command and control structure, it might have determined that committing a crew to perform a rescue on the upper levels of the building with a developing fire below was not a viable option (not an acceptable risk) – at least until the vent on the mezzanine floor through the open door was controlled - in terms of satisfying the objective of safeguarding firefighters. However, what is known is that this mode of operation is routine in the Hellenic Fire Service and taking place every day. A very similar incident occurred recently in Athens, where a firefighter and officer became trapped on the top floor of a block of flats and could not descend due to the heat of emitting from the fire via an open door on the lower floors. On this occasion, both firefighters were lucky and managed to escape onto the roof by breaking a small window at the top of the stair.

Communications

In 2004, most of the emergency services in the prefecture of Attica acquired a Terrestrial Trunked Radio system (TETRA), leaving the old analogue VHF system as a back-up resource (though the analogue system was still maintained outside Attica). This communications system came about to provide greater assurance and security during the Athens Olympic Games in 2004. The service provider for this system was Siemens-SAIC and the network was set-up and supported through the National Communications Organisation, OTE. Though this system promised a number of services that were to modernise the Hellenic Fire Service's response and resource management systems by incorporating Global Positioning System (GPS) monitoring of fire appliances combined with Geographic Information System (GIS) capabilities to assist with co-ordinating resources, especially at forest-fires, only a very small part of this system was ever handed-over to the Hellenic Fire Service. This meant that the only added value it offered was an enhanced level of security to radio transmissions as the digitalised encryption could not be monitored by external sources or people with scanners.

Furthermore, all radio terminals had personal identification numbers and the control room was able to identify any user pressing the Press-to-Talk (PTT) button. This feature alleviated outside interference on active user groups (equivalent of radio channels), something that was always a problem on the analogue system where members of the public could scan and listen-in to emergency services communications and interfere with transmissions. It also minimised horseplay over the radio seeing as any user could be identified almost instantly through their terminal's unique identification number. This system had been provided on a 10-year contract starting on 1st September 2004 and ending on 31st July 2014 with an option for the Greek government to either extend the contract for a further 5 years or receive complete handover of the system, thus, assuming full responsibility for its maintenance and servicing. The TETRA system was procured at a total cost of €114,000,000 with a contractual commitment for servicing and maintenance of an additional €82,000,000. However, things did not happen as had initially been agreed with the TETRA system presenting numerous problems from the outset (Kafantari, 2014).

One of the key issues was that, because this system had only been implemented in Attica, in the event of large fires where resources would be drawn from other prefectures, radio communications required a two-tiered approach as there was no interoperability between the old and new systems. This significantly complicated incident management and was inherently dangerous for operational crews, especially those operating on the analogue system. Most volunteer organisations in Attica that supported the fire service, especially in the early years (2004 – 2010), were only provided with one portable terminal per fire station (if any) meaning that in multiple appliance stations, many units would need to respond to fires using the analogue radio system. The incident would initially be dispatched on the TETRA system and many units would then respond to the incident utilising the analogue radio without any capability of listening-in to incident updates or the ability to communicate with other appliances that were using TETRA. In most cases, it was considered pertinent and, indeed, a necessity for the station to maintain the TETRA terminal to monitor any other incidents occurring in the local areas. This is because the main dispatch channels were on the TETRA system and the analogue channels were generally discarded and only utilised as a back-up resource by the control room officers.

In the significant forest-fire of North-East Attica in August 2009 that lasted several days, the author recalls that on the second day of the fire, he and his crew became trapped trying to save a local resident who refused to leave his home in the area of Stamata[42]. Multiple calls for assistance over the analogue radio went unanswered by the control room and narrow escape was made in the safety of a nearby vineyard. However, one full-time fire appliance with the driver alone did not manage retreat and it was impossible to determine whether the driver was safe until the fire had passed through. Though the driver had literally driven through the fire into an open plot of land, it was only via a telephone call to the control room that this critical information could be divulged. It was also common practice that the control room officers coordinating resources on the TETRA system would turn-down the analogue radio as it served as a distraction when trying to coordinate resources over the main TETRA system. This is a reasonable solution to a problem created by the lack of consistency in radio communications and lack of appropriate resources to manage two separate systems for the same incident. Logistically, this did not make sense and the inadequacy of this approach was demonstrated in practice time after time.

---

[42] Stamata is a local town of the municipality of Dionysos in the North-East of Attica.

In reality, it is estimated that the TETRA system additionally cost the Greek taxpayer over €120 million in the 10-year period that it was utilised (Kafantari, 2014). This contract formed part of the Siemens C4i scandal, one of the largest scandals uncovered in the period following the financial crisis in 2008 which saw government officials and other domestic and foreign executives for Siemens being charged with offences of bribery, embezzlement and money laundering in order to secure contracts for telecommunications products and services from Siemens-SAIC from the late 1990s (Associated Press, 2008). As a result, when the contract finally came to an end, the Greek government had no choice but to return to the old analogue system that was outdated and poorly maintained. Though some improvements have been made over time, in 2021 there are still areas in Athens where the radio system has inadequate to non-existent coverage. There were areas in the northern suburbs of Athens where communication with the fire service control room can only be made through mobile telephones. This became particularly complicated where mobile phones also had no reception.

On the night that Mantzios died, the fire service had since converted back to the old analogue system and the incident was being managed on channel 1. This channel serves as the main dispatch channel for all metropolitan fire stations in and around Athens and is also used as the main incident channel (if the incident has been dispatched on this channel) i.e., this channel serves two purposes. It is quite rare for any incident to be managed on a different channel in Greece other than on its main dispatch channel though in some extreme cases like in a number of serious riots or flash-floods in Athens, initiative has been taken and this has occurred. It must also be stated that no written protocol for radio-communications was identified as part of this research, though there are some rules that are learned, passed-down and generally adhered to. At least in the case of volunteer fire stations, no radio communications training is provided. Most firefighters learn effective radio communications through practice, learning how to listen and repeating or mimicking sequences or phraseology that is considered as appropriate and more formal. As such, radio communications are very much an experienced-based process. Control room officers often need to restore radio discipline and will probe operational response crews to acquire the correct information so that appropriate management of the incident can then be undertaken. Experienced control-room officers are very good at policing radio-discipline and acquiring the required levels of information from operational crews working at an incident. At large incidents, there are many occasions when radio communications are chaotic and where it is almost impossible to send or receive a message without 'stepping-over' someone[43]. This creates a level of unreliability in the communication chain.

Mantzios was the only firefighter of his crew on the night of the fire equipped with a hand-held radio. What is particularly interesting is that in most of the case-studies from the United States described earlier, a recurring theme was that though the control-room did not hear firefighter's calls for help, members of the public or persons remote to the incident did. This was also the case in the Mantzios incident. Though repeated calls for help were sent through to the control room, it appears that these were not heard or not understood. It was only following the intervention of a patrol-unit from the 1st fire station of Athens that the control-room appeared to fully understand the message and the severity of the situation. It is also known that numerous other phone calls were made to the control room by other fire patrol units, full-time and volunteer fire stations, members of the public with scanners and off-duty personnel to report the situation and the messages sent through by Mantzios

[43] 'Stepping over' is when multiple persons are using the press-to-talk function on the radio at the same time leading to none of the communications going through to their desired recipient.

requesting for assistance. Indeed, this issue appears to be a contributing factor in many fatal fires. It is a known issue and one that most probably has to do with the configuration of the system itself.

However, what is also particularly noteworthy is that not one of Mantzios' crew members reported hearing his calls for assistance. Fire appliances in Greece are fitted with a facility that sends the receiver speaker from the radio from the front cab of the fire appliance to the back in the pump-bay area, so that the driver and others in the vicinity can hear radio messages. Apart from the fact that no other member of Mantzios crew was equipped with a personal hand-held radio, the one driver was assisting firefighting crews deal with the fire in the shop on the ground floor and the other was assisting the contract firefighter, that was originally with Mantzios, in making rescue attempts to re-enter the structure via the main staircase. As such, no-one was listening to the radio and no-one at the fire scene heard Mantzios calling for help. In any other situation and in many other countries, all firefighters are equipped with their own hand-held radios or at least have one radio between a team of two. The Incident Commander is responsible for monitoring radio traffic at the incident and managing resources appropriately. These responsibilities might extend to sector Commanders who will manage smaller areas of an incident and report into a more senior Incident Commander overseeing the entire incident. The point is to ensure accountability of crews operating in risk areas with forethought around their safety and welfare.

At this incident, it is apparent that in the early stages and before further resources arrived there was no formal accountability process for anyone. When reading through the internal incident report and from discussions with firefighters who were listening to the incident from start to finish, certain levels of accountability appear to be acquired by virtue of the personnel's stated or assigned positions at the fire. However, this process appears to lack formality and defined structure and serves as an ad-hoc function to the overall incident without the required levels of granularity. A formal accountability process should be able to identify who is working where, for how long and what tasks they have been assigned. Where line of site is not possible, the only way of ensuring this accountability chain is through radio communications, where updates on a crews status can be given to report any changing or dangerous circumstances (Trippett, 2013). Without these processes, firefighter accountability cannot be ensured and their exposure to risk is likely to increase significantly as reasonably foreseeable risks will not be identified, communicated and mitigated early on. In Greece, there is still no formal accountability process for operational incidents, and this is broadly recognised as a key contributing factor to firefighter fatalities on a global level (Parks, 2008; Angulo, 2010). Though, in most firefighter fatality case-studies, the lack of radio communications alone did not serve as the sole factor leading to the fatalities, it did serve as a significant contributing factor in an overwhelming number of cases. To date, in Greece, firefighters are still not provided with personal hand-held radios. Many of them, in understanding their value and importance for their personal safety, have purchased their own radios. This is not an ideal solution, as the quality of the radios cannot be guaranteed and the programming and / or the servicing that they might require are unlikely to be undertaken adequately by the individual firefighters.

Health and Safety in the Hellenic Fire Service

It was not until 2003 that work-related accidents and illnesses were incorporated into legislation for Greek firefighters. Work related health and safety legislation was first published under law 1568 in 1985 but did not apply to the public sector, especially the emergency services, until the application of law 3144 in 2003.

This meant that until that time, any injury or illness was not associated to the workplace and, as a result, firefighters were not eligible for equivalent compensation and treatment as applied for workers in the private sector. This also meant that injuries and illnesses were not recorded or reported at an organisational level. (Chalaris, 2013: p 1). However, law 3144 only applied to their place of work and not to operational incidents, effectively meaning that incident sites were not considered as places of work. All aspects relating to the application of health and safety law to fire service operations were meant to be covered under separate regulations in the form of Presidential Orders. Until 2019, these regulatory documents had still not been put into force.

In March 2018, the United Firefighters Movement in Greece organised a health and safety conference in the north of Greece in the city of Thessaloniki to discuss the critical issues around health and safety in the Hellenic Fire Service. On behalf of the National Federation of Fire Service Employee Unions, Dimitrios Vlachos reported the following issues (Rizospastis, 2018):

- The upper limit (working hours) that firefighters can safely operate at an incident has not been identified or consolidated. This means that there is no employee protection against unreasonable or unsafe working hours. As a result, employees are subjected to immense strain from the point that they can often work 30 to 40 hours continuously, especially at forest fires where their relief is highly dependent on the size and intensity of the incident. Such incidences are further exacerbated by the decline in full-time staff.

- The required and modern personal protective equipment has not been issued as appropriate and there are significant problems in the issuance in even the most basic of personal protective equipment to 5-year contract and seasonal firefighters[44].

- There is a wholly inadequate system around the logistics relating to the provisions of food and water for firefighters operating at large incidents.

- The fires service is incapable of securing adequate accommodation for firefighters when they are operating away from their local region at large sale incidents. On many occasions, they are advised to reside at military barracks, as they have done in the past on islands such as Rhodes and Chios, where the conditions are not appropriate. As such, many firefighters are forced to rent hotel rooms which they pay for themselves in order to have the luxury of a shower, a hot meal and a good night's sleep. On other occasions, firefighters might be sent to rest at other regional fire stations which causes problems for the local staff and their stations are not designed to facilitate such large number of firefighters. Alternately, firefighters might just rest (sleep) inside the fire trucks or outside on the ground.

These conditions are, indeed, a realistic depiction of the conditions faced by Greek firefighters. The author has attended many large incidents where resources have been sent from all over Greece and the conditions are precisely those described above. This is a big difference from other fire services in Europe that have better management procedures around firefighter

[44] It is interesting that there is no mention of the significant problems with the issuing of personal protective equipment to volunteer firefighters. Despite fulfilling significant operational roles at full-time fire stations across the country, personal protective equipment has never been issued to volunteer firefighters. Most of them will either have to purchase it themselves (new or second hand), acquire it through donations or it is handed down from older firefighters ready for retirement.

health and safety, allowable upper limits for working hours and provisions for food, drink and rest. At the conference, health and safety specialist, Dr Evi Georgiadou (2018), pointed out the risks that are presented to the public if firefighter health and safety is compromised. She stated that the work of firefighters is such, that if their health and safety is not protected, then the public's health and safety will not be protected either. She added that, many firefighter accidents could have been prevented. Georgiadou emphasised that the fire service does not have a steady work environment, contrarily, it is constantly changing, with multiple dangers, multiple responsibilities, and the nature of rapid intervention for every incident increases the risk exponentially. Additionally, she added that the overexertion and strain experienced by firefighters in the line of duty is a significant contributing factor in firefighter fatalities arising from heart conditions or even suicide.

The research around the health risks that firefighters are exposed to as a result of contaminants has become quite prominent in the last few years. Many fire services have adopted measures to minimise the exposure of firefighters to pollutants and other contaminants, especially those that saturate or remain on their PPE after an incident. As discussed earlier, it is the exposure to contaminants that remain on the uniform and come into direct contact with the skin, respiratory system etc. that create higher risks for firefighter illness. Wash-down procedures after an incident, the provision of anti-bacterial wipes, industrial washing machines at fire stations and CO_2 decontamination, bagging and placement of dirty PPE into bins, the restriction of dirty PPE being worn in fire trucks, the allocation of clean and dirty areas at fire stations and regular health screening are all processes undertaken by fire services to minimise the exposure of firefighters to contaminants that could lead to serious health risks. This is not the case in Greece.

A firefighter from the Sixth Fire Station of Athens, Vasilis Giannakos wrote an article titled 'Contaminated PPE: Instead of Protecting they Kill' for the National Fire Service Magazine 'Fire Service Inspection' in 2019. In his article he identified the risks presented to firefighters from contaminated PPE and the measures that are required to mitigate these risks. He then explained that, unfortunately, in Greece, these concepts are not yet understood. He argues that though there are washing machines at many fire stations, it is common practice for firefighters to take soiled PPE home and wash it in the domestic washing machine as well as rest, sleep, cook and eat in it as well, even directly after an incident. He explains that the increase in firefighter cancers is unavoidable but unfortunately there is no such data recorded by the fire service in Greece to quantify this statement. Most importantly, however, Giannakos makes reference to the poor culture around health and safety in the Hellenic Fire Service that cannot only be regulated by legislation. Giannakos argues that personal dedication and understanding around health and safety are imperative.

A big problem is the absence of health data for firefighters. Though there is a memorial to firefighters that died in the line of duty on their internet webpage[45], there is no analysis of the details around their death where these are related to acute health conditions such as heart-attacks. There is a general categorization inferring a health consequence rather than an injury, and, furthermore, there has not been one 'lessons-learned' study published from any firefighter fatality. Since 1942, 56 firefighters have died in the line of duty in Greece. The number of firefighters that have died from work-related health conditions such as cancers, lung disease, heart attacks and strokes etc. are not recorded. In the table below are the Line of Duty deaths recorded from 1980 onwards. These are categorized as follows:

[45] https://www.fireservice.gr/el_GR/pesontes-sto-kathekon1

Number	Year	Cause of Death
1	1982	Burns
2	1989	Heart-attack
3	1991	Road Traffic Collision
4	1996	Drowning
5	1997	Road Traffic Collision
6	1997	Road Traffic Collision
7	1998	Roof Collapse at Fire
8	1998	Injury during Forest Fire
9	1998	Injury during Forest Fire
10	1998	Injury during Forest Fire
11	1998	Injury during Forest Fire
12	1999	Boiling Liquid Expanding Vapour Explosion – Propane Tanker Fire
13	1999	Boiling Liquid Expanding Vapour Explosion – Propane Tanker Fire
14	1999	Boiling Liquid Expanding Vapour Explosion – Propane Tanker Fire
15	1999	Road Traffic Collision
16	1999	Road Traffic Collision
17	1999	Injury during Forest Fire
18	1999	Injury during Forest Fire
19	2000	Injury at forest-fire leading to death 37 days later
20	2001	Structural fire incident
21	2001	Recorded as 'unknown', occurred while refueling fire truck
22	2003	Road Traffic Collision
23	2005	Fatal Injury at structural fire incident
24	2007	Injury during Forest Fire
25	2007	Injury during Forest Fire
26	2007	Injury during Forest Fire
27	2007	Injury during Forest Fire
28	2007	Injury during Forest Fire
29	2007	Injury during Forest Fire
30	2007	Injury during Forest Fire

31	2007	Injury during Forest Fire
32	2007	Injury during Forest Fire
33	2011	Road Traffic Collision
34	2011	Injury during Forest Fire
35	2011	Road Traffic Collision
36	2012	Sudden illness during forest fire
37	2013	Road Traffic Collision
39	2014	Asphyxiation at Structural Fire
40	2016	Sudden illness during training
41	2016	No cause given – assumed sudden illness whilst on shift
42	2016	Sudden illness during forest fire
43	2017	Injury during Forest Fire
44	2018	Injury during grass fire
45	2019	Injury during grass fire
46	2020	Sudden illness during forest fire
47	2020	Sudden illness at incident
48	2021	Drowning
49	2021	Struck by car
50	2021	Sudden illness during forest fire
51	2023	Sudden illness during forest fire

Table 5 – Firefighter fatalities in Greece from 1982 – 2023.

In this data, it is interesting to see the increase in sudden illness events from 2012 onwards. These are assumed to be incidents relating mainly to sudden cardiac arrest. Though there is anecdotal evidence around this, there is little formal information published to confirm it. It is unknown whether there is a genuine increase in such incidents or whether these were simply recorded. These firefighters, however, were between 30 and 50 years old when they died. The author is aware of at least another four cases since 2000 where firefighters have suffered heart-attacks or strokes whilst on shift or directly afterwards, the majority of whom were between 30 and 50 years old. He is also aware of at least six cases where firefighters have been diagnosed with or died from various cancers. This information is not recorded or not publicly available. The Hellenic Fire Service publishes annual performance reports. In these reports they record injuries to firefighters and members of the public. However, this information is only quantitative without any qualitative explanation as towards the causal factors leading to those injuries. In 2018, the Hellenic Fire Service reported 17 firefighter injuries, 2 burn injuries and 1 firefighter fatality that occurred at a wildland fire. Conversely in 2017, there were 11 recorded firefighter injuries and 1 fatality. In 2016, there was a total of 14 firefighter injuries and 1 burn injury recorded.

Perhaps the most critical element in respect of health and safety is with the regard to the extent that it is prioritised at organisational level and forming part of fire service culture. Across the world, most fire services understand the need to regulate health and safety practices through policy and procedures. Even in Greece, a study undertaken by Professors Gousopoulou and Hourdaki of the Dimokriteion University of Thrace in 1994 around Health and Safety in the Hellenic Fire Service identified that in order to reduce the likelihood and frequency of firefighter accidents there needs to be a robust standard of training and the creation of standard operating procedures that incorporate safety regulations at operational incidents. Hellenic Fire Service Officer, Michael Chalaris (2013: pp 11 – 12) also asserts that the investigation of accidents will further assist in reducing injuries. These are perhaps the most salient points in the Mantzios case study fire.

SOPs are not developed to only provide tactical guidance for dealing with incidents. They incorporate safety considerations around foreseeable risks, but fire firefighters and fire officers are also trained in the undertaking of dynamic risk assessments that will ensure that they can consistently operate in a safe manner. The job of a firefighter has inherent risks, but these are compounded where procedures are not in place to regulate firefighter actions. If one were to take the example of the first entry made by Mantzios and his colleague into the building, it was established immediately that the stair was heavily smoke logged. This is a vital piece of information that needs to be risk-assessed. In a system with a robust incident command structure, this information would have been fed back to the incident commander who would have evaluated it through a decision-making process and dynamic risk assessment. They would likely have tried to establish the location from where smoke was entering the stair and based on their operational guidelines to not work above a fire if it is not controlled or contained. Sadly, once the Mayday was called by Mantzios, there was no formal procedure for dealing with. The control room did not initially understand the message, no-one on the incident ground heard his calls for assistance. An aerial appliance was not mobilised in the initial stages of the fire even though there were reports of persons trapped on the upper levels of the building. The deployment of an aerial appliance appears to be standard procedure for the European fire services that commented on this case study. The additional resources that were sent were mobilised too far into the incident to have made any difference to the outcome. All the attempts that were made to rescue Mantzios were reactive and undertaken through the central stair that was effectively acting as an exhaust vent for the fire. A rescue via the central staircase would have been similar to entering a chimney with the fire lit in the fireplace below. There was further no attempt to utilise other equipment at the scene of the incident like the 9m ladders that on-scene fire trucks were equipped with. The international fire professionals that responded to the case study exercise all commented to the author that this fatality was, in their view, 100% avoidable.

It is not the intention of this book to criticise the firefighters involved in this incident, especially those that made consistent efforts to rescue Mantzios. It is clear that all those involved did what was within their powers to save him. Furthermore, as the author was not present, the intricate conditions that were experienced at the fire scene might not be entirely understood, except where this information has been provided as evidence by firefighter statements in the incident investigation report. It is believed that no individual firefighter can be held accountable for any failures in this incident, especially Mantzios. It is thus purported that the failures at this incident are fundamentally of a systemic nature. The firefighters did not operate outside any guidelines as there are none to work to in the first place. The level of training is poor and most fire services in the world would not have permitted a crew like this to respond to an emergency incident.

The control room did not operate any differently than any other incident as, again, there is no guideline to determine response configurations, and these are largely based on reactive and intuitive criteria. Mantzios actions were no different than any other firefighter operating at similar incidents every day in Greece. It is common practice for incident commanders to undertake operational hands-on activities and work above the fire without the fire below being controlled. There is no understand or guidance around not placing oneself in the flow-path of the fire (the area between the fire and an exhaust vent) and most of the time, actions relating to firefighting, ventilation and rescue activities are not coordinated with each other, which puts firefighters and civilians at significant risk. There is a poor organisational culture around the implementation of formal processes due to the absence of policies and procedures, and the fact that no 'lessons-learned' reports arise from such events means that the same mistakes are repeated over and over again. In this case, the systemic failures killed Mantzios and nothing seems to have been learned at organisational level from this tragedy.

The next chapter explains the methodological approach used for data acquisition and its analysis. The purpose of this investigation was to test hypotheses regarding perceived differences between south and north European fire services. This research commenced with a preconceived opinion that fire services in south Europe were not as efficient, effective or well trained as north European firefighters and that an absence of processes creates opportunities for freelancing behaviour at emergency incidents, therefore increasing the risk of a negative outcome. It was also theorised that due to the absence of sufficient funding that fire services in southern Europe were less likely to develop in comparison with their north European counterparts.

Chapter 5 – Research Methodology

There are several stages to the research approach of this study. The different research styles are aimed at addressing different parts of this study. The analytical framework employed in this research is underpinned by a convergent mixed-methods design, integrating both qualitative and quantitative methodologies to produce a comprehensive, multi-dimensional understanding of the intrinsic issues within fire and rescue services that will ultimately affect the quality of service they provide to the communities they serve. This approach enables the interrogation and cross-examination of diverse data sources, providing a robust and comprehensive analysis of the factors affecting fire service governance and operational safety across Europe due to the absence of baseline standards.

Qualitative Study

At the core of the research lies a case study analysis, with the Mantzios incident serving as a critical case that identifies broader implications to fire services with weak organisational structures. It should be iterated that the Mantzios case study informed, and indeed instigated, the general concepts and questions of this book. Although more than a decade has passed from this incident, there are still recent examples of near-miss incidents in the Hellenic Fire Service resulting from the same mechanisms of failure. The analysis of this case study was undertaken by via a series methods. The first was via the acquisition of the internal fire service investigation report into the death of Mathaios Mantzios. Additionally, the author acquired the technical report drafted on behalf of Mantzios' family by Adrianos Gourbatsis.

To achieve greater credibility and impartiality in the analysis of this incident, the author implemented a three-stage approach.

The first stage was an in-depth literature review of the historical and contemporary problems experienced by fire services on a global scale. Several other case studies into firefighter fatalities were reviewed such as the King's Cross fire in 1987, a shop fire in Bethnal Green in 2004, a high-rise flat fire in Shirley Towers, Southampton in 2010, a fire that occurred in Belgium in August 2019 where two firefighters lost their lives, and numerous fatal fire incidents in the U.S.A. Other government, national and international firefighter injury and fatality data were analysed extensively, along with further literature, guidance documents, technical reports, articles and policies concerned with fire service operations, organisational culture, and the development and modernization of fire services. This helped to identify and categorize the systemic failures identified in Mantzios case study and pertinent to the Hellenic Fire Service, which were analysed against a range of basic principles that were drawn from the literature review.

The second stage was based on a case-study exercise with a series of participants. A redacted version of the Mantzios fire incident was sent to several fire service experts from northern Europe, Spain and Australia. The purpose of this exercise was to establish the different approaches that other fire services would have adopted in dealing with this incident based on the information provided and to determine whether these decisions would be governed or informed by a standardised procedure or set of guidelines. It was also important to understand the logic to each approach and the strategic rationale behind them. In addition, the aim of the exercise was to establish commonalities or differences between fire service responses in order to identify patterns that would demonstrate inefficiencies of the response configuration sent to this incident by the Hellenic Fire Service.

A second element of this exercise was to send the full case-study to a selection of the participants that took part in the first part of the exercise and then undertake a discussion in respect of their personal views on the errors that occurred during this incident in Athens. A number of other participants were sent the full-case study directly without being included in the first stage of this exercise to counter any bias that might have been influenced by the primary incident information. These individuals included retired fire officers, fire service specialists, professors and fire safety engineers renowned in the field as contributing significantly to the industry. In total, 14 fire professionals were engaged from Belgium, the Netherlands, Germany, Spain, Portugal, Greece, Australia and the UK.

The interviews were held via face-to-face meetings on Skype and some participants chose to provide their responses via email. The interactive interview questions were open ended and mainly allowed the participants to develop a working hypothesis about this incident through open discussion. Occasionally, they would ask specific questions relating to the incident, the Hellenic Fire Service, its culture, training standards, the existence of procedures and so on. The responses that were provided were factual but due to the absence of formal processes and procedures, some responses required input based on the author's personal experience of how the fire service in Greece operates. These opinions tend to be broadly shared across the sector in Greece and some aspects have been documented in articles and reports about the fire service by independent individuals, journalists, NGOs and fire service union representatives.

It was established in the latter stage of research that there was a generalised absence of consensus on a global scale between fire services, meaning a set of baseline standards relating to the efficiency and effectiveness of fire services does not currently exist. Therefore, the literature review was extended into much broader areas in order to understand the strengths, weaknesses and perhaps to debunk some misconceptions about the fire service, especially with regard to those fire services that are perceived as world leaders. It was important to understand this context holistically.

Quantitative Study

Complementing the qualitative findings, descriptive statistical analysis was applied to a large-scale questionnaire distributed to 576 firefighters across Southern and Northern Europe. This research approach was structured around key thematic areas identified in the Mantzios case study as contributing factors (systemic failures) that led to his death. These were identified as contributing factors, not only in the Mantzios case study, but also in other fire service line-of-duty deaths studied as part of this study, thus linking the applicability of mechanisms of failure between fire services. These thematic areas formulated categories under which a series of questions were asked of the respondents. The thematic areas include:

- Personal protective equipment
- Training
- Communications
- Incident command
- Resource management and health and safety

Participant demographics were also requested in order to establish age, gender, years of service and rank. Participant demographics would inevitably assist in identifying differences of opinion influenced by factors such as years of experience. The questionnaires were designed on Google Forms and aimed to acquire the respondents' personal views on matters relating to the thematic categories stated above, but also to establish an understanding of the general structure and problems relevant to their specific fire service.

The questionnaires were translated into Greek, Portuguese, Italian and Spanish and sent out via email link that could be opened and submitted online anonymously and were subject to a compulsory consent form that was required before persons could move onto the questions area. The questionnaires were sent to firefighters who assisted in translating them into their own language, thus ensuring that specific terminology and details relating to their fire services were not overlooked. Respondents had to answer all questions in order to complete and submit the questionnaire. The same questionnaire was also created in English and sent out to UK fire and rescue services which were established as the control sample. The reason for choosing the UK fire services as the control sample was because of the ease of access to information, which is publicly available and, of course, due to the language. Despite the significant changes arising from the Fire and Rescue Services Act in 2004, which effectively decentralised many of its processes and procedures, the UK Fire and Rescue Service still has a reasonably structured approach on a national level in a number of the thematic areas mentioned above, and the systems are simple to understand and analyse. The UK FRSs also operate more in-line with north European fire services than do those in the United States or Canada, for example.

Information on UK fire and rescue services is extensive and transparent. There is also a reasonable balance between political and strategic position, as fire service related topics are covered internally by fire services themselves, by academics and researchers, the fire brigades union, and by independent auditing organisations such as His Majesty's Inspectorate of Constabulary and Fire and Rescue Services (HMICFRS). The data from these questionnaires would additionally result in an effective comparison between the fire services in southern Europe and the control sample.

The Greek, and Portuguese questionnaires have been translated and completed, gathering just over 200 responses from the Hellenic Fire Service and just under 200 from the Portuguese fire services. Some issues that have been encountered at this stage of the research are in respect of language and the organisational context of each fire service. It was discovered that in some cases, questions required rephrasing between languages to capture the context of the question. This was also pertinent between fire services that had slightly different structures and processes. Some examples included rank structure that were notably different between countries. Though an attempt was initially made to group these in officer and non-officer roles, the differences between fire services were too vast and, therefore, the respondents were evaluated as firefighters collectively and differences were evaluated against other parameters such as age, experience and gender.

Another notable difference was between the terminology used to describe volunteers. Whereas Greece, for example, utilises unpaid firefighters, those described as volunteer firefighters in northern Europe are in most cases part-time or otherwise on-call firefighters. This category of firefighter is on standby at their place of business or at home and alerted via electronic pager when required to attend an incident. From the point that they are alerted, these firefighters are paid on an hourly rate. This is very similar in the UK where most fire services employ Retained Duty System firefighters who are paid a retainer fee, an hourly rate, a call-out fee, and their time spent at weekly training nights. It is evident that this category of firefighter is not a volunteer by virtue of their services being paid. However, it was found that unpaid volunteers and on-call firefighters are generally held in the same regard by full-time paid firefighters. Due to the differences in their structure and utilisations, the sections relating to volunteers specifically was removed from the analysis section although volunteers have responses in an operational capacity as fire service staff and included within the dataset as firefighter respondents.

There was also significant difficulty in collecting responses from UK fire fighters. Despite contacting organisations directly as well as the Fire Brigade Union, the National Fire Chief's Council and His Majesty's Inspectorate for Constabularies and Fire and Rescue Services (HMICFRS) there has been very little or no feedback at all, with many communications going unanswered. It total, only 45 responses were received from UK FRSs. This stage of the research used social media platforms (such as Facebook) as well as emails and word of mouth to contact respondents to disseminate the questionnaire.

The absence of data is not necessarily a problem in the case of the UK, as National Occupational Standards will generally serve as the principle organisational structure for FRSs around matters covered in the thematic areas of this questionnaire. This is evident from the responses that, though not extensive, have many similarities despite wide geographical representation. Similarities were also identified in the responses provided by firefighters from Northern Europe. As such, the UK responses and north European responses have also been grouped to allow for easier evaluation between northern Europe (including the UK) and southern Europe.

In addition, though many attempts were made, it was not possible to acquire sufficient data from Italian firefighters. The fire service in Italy is a national service except in the Region of Valle d'Aosta, and Bolzano & Trento Provinces (Civil Protection Department of Italy, 2018). Although there was a fair geographical representation among the respondents, it was not felt that these data were reliable enough to pass sound judgement on the structure of the Italian fire services. Nevertheless, the data will be used for broad comparisons, but the reader should be aware of the reliability constraints of the data.

The purpose of this quantitative exercise was to establish specific patterns and opinions that would demonstrate a clear differentiation between fire services in the two halves of the continent. Quantitative data were analysed using frequency distributions and cross-tabulations, enabling the identification of regional disparities and validating key patterns observed in the qualitative analysis. While the current analysis is primarily descriptive in nature, the research design lays the groundwork for potential inferential statistical testing in future studies. For instance, chi-square tests could be employed to assess the strength of associations between geographical location and variables such as firefighter satisfaction with training or confidence in SOPs.

In summary, the data analysis strategy employed in this study reflects a methodologically rigorous and context-sensitive approach. The integration of comparative case analyses, thematic interpretation, and descriptive statistics enables a nuanced exploration of both individual and systemic factors contributing to operational risks in the fire service. As a volunteer firefighter, the author possesses first-hand experience of fire service operations and broader matters relating to organisational culture of fire services etc. This is not only specific to the Hellenic Fire Service due to the author's affiliation to international firefighter associations, focus groups and training initiatives abroad that give him a broader understanding of the problems experiences by fire services contemporarily. While this background has provided the author with valuable insights into the realities of frontline work, he has remained conscious of the potential for personal bias and how this might influence the analysis and interpretation of data and other findings.

To address this, the author has employed a range of strategies designed to ensure objectivity and academic rigour. A reflexive journal has been maintained throughout the research process to document and critically reflect on any assumptions, decisions, and emotional responses. Methodological triangulation has also been applied, drawing on case study analysis, expert consultation, and survey data, to cross-verify findings from multiple sources. Peer debriefing was an integral part of this process, with academic supervisors and professionals from outside the author's immediate professional network providing independent feedback and critique. The questions for the questionnaire were presented in neutral language, without a leading context and did not presume a response that might have influenced the language of direction of the question from the outset. Through these combined strategies, the author has aimed to uphold transparency, enhance the credibility of the research, and ensure that the findings are grounded in systemic evidence rather than individual experience.

This research adhered to fundamental ethical principles including respect for persons, beneficence, and justice. Ethical risks were minimal, and every effort was made to ensure that the research was conducted responsibly, with integrity, and with respect for participants' dignity and confidentiality. By incorporating culturally sensitive methods, ensuring informed and voluntary participation, and maintaining a high standard of data security and transparency, the research met the expected standards of good ethical practice.

Chapter 6 - Data analysis and Results

In total, 576 firefighters responded to the questionnaire. The majority of respondents (206) were firefighters from Greece, and they represented 35.8% of the total number of respondents. The second largest number was from Portugal where 186 responses were received, constituting 32.3% of the total number. Spanish firefighters provided 72 (12.5%) responses, UK firefighters 45 responses (7.8%), 35 (6.08%) responses came from Italian firefighters and 32 responses from firefighters in North European fire services such as Germany, Belgium and the Netherlands.

Response Rates	Count of Timestamp	Count of Timestamp
Greece	206	35.76%
International	32	5.56%
Italy	35	6.08%
Portugal	186	32.29%
Spain	72	12.50%
UK	45	7.81%
Grand Total	**576**	**100.00%**

Table 6 – Response Data by Country

Respondents by gender was predominantly male (89.2%) against only 10.8% of women. However, this ratio reflected the staffing patterns of most fire services that openly publish staffing statistics. Some examples are shown below:

Country	Percentage of Female Firefighters
United States (2019)[46]	8%
England (2021 Census)[47]	7.5%
Greece[48]	5%

Table 7 – Percentages of Female Firefighters

Out of the 576 respondents, the majority were aged between 34 and 41 (29.5%) and then 42 to 49 (28.5%), followed by the age group 26 – 33 (16.7%), then 50+ (12.8%) and lastly the age group 18 – 25 (11.6%). Similarly, the majority of respondents (37.5%) stated to have between 10 and 19 years of service followed by those having between 20 – 29 years of service (21.9%) and then those with between 5 and 9 years of service (15.4%).

Personal Protective Equipment

It is understandable that in order for a firefighter to operate efficiently and safely, they must be provided with appropriate personal protective equipment (PPE). In addition, measures should also be taken to ensure that firefighters are not exposed to further risks post-incident such as exposure to carcinogens from toxic fire effluents found on contaminated firefighting equipment. To mitigate such risks, further processes need to be in place to clean, decontaminate or replace dirty or damaged

[46] https://www.nfpa.org/News-and-Research/Data-research-and-tools/Emergency-Responders/US-fire-department-profile

[47] https://www.gov.uk/government/statistics/fire-and-rescue-workforce-and-pensions-statistics-england-april-2020-to-march-2021/fire-and-rescue-workforce-and-pensions-statistics-england-april-2020-to-march-2021

[48] https://cityportal.gr/ellinides-pyrosvestries-gynaikes-kataktoyn-androkratoymenoys-choroys/

equipment, further supported by stringent guidelines for firefighters' personal hygiene at organisational level (Stec, 2020). The first point to determine was whether firefighters felt that their PPE is suitable for the incidents they attend. In the grouped sample for north European firefighters, (including UK firefighters) 82% agree that their PPE is fit for purpose. Conversely, 63% of firefighters in the sample group for south European firefighters feel that their PPE is suitable for the incidents they attend. The Portuguese (37.6%) and Greek (34.5%) firefighters presented the highest levels of dissatisfaction with the suitability of PPE provided by their respective fire services. In the case of Portugal, it is known that PPE used by voluntary fire services is often either purchased or donated second hand from north European fire services[49].

Similarly, in Greece, many full-time firefighters have only ever been issued one set of PPE which they have retained well beyond its safe use date. There are no standard procedures for cleaning and maintenance or health and safety measures for stowage and decontamination. This means that many firefighters often take soiled PPE home and wash it in domestic washing machines along with other civilian clothes, which could in many cases be of family members i.e. children, spouses or partners (Giannakos, 2019: pg-7). Volunteer firefighters are not provided with PPE leading them to purchase their own equipment which is often of questionable suitability (Pantelides & Tziritis, 2009: p - 16).

There are notable differences between the two grouped models, north and south European fire services, with 70% of respondents from north European fire services agreeing that they felt that they have been provided with appropriate PPE against only 45% of firefighters from south European fire services. The most notable differences between countries are

---

[49] https://eu.beaconjournal.com/story/news/nation-world/2017/08/15/heroes-without-capes-portugal-s/10770790007/

between the United Kingdom and Greece where 77% of firefighters in the UK feel that they have been provided with appropriate PPE, whereas only 17.5% of Greek firefighters feel that they been provided with appropriate PPE.

Conversely, when asked if firefighters have ever had to purchase their own PPE, 27.3% of the grouped sample of north European firefighters (including UK firefighters) stated 'yes' whereas 65% of the grouped south European firefighters said that they had purchased PPE. The most notable differences for this question were again between the UK and Greece where only 17.8% of respondents for UK fire services attested to buying their own equipment against a notable 96% of Greek firefighter respondents. 54% of Portuguese firefighters reported buying their own equipment and approximately 22 and 25 % of firefighters in Spain and Italy.

The main reasons firefighters bought their own PPE were presented in the three following phrases of which firefighters could choose only one response:

- *"I bought my own equipment in order to ensure my safety as my fire service will not provide me with this equipment or will not replace lost or damaged equipment."*

- *"I bought my own personal protective equipment to add or enhance what my fire service has already provided me with".*

- *"This doesn't apply to me".*

In the grouped samples, north European firefighters (including UK firefighters) represented 9% of the responses relating to purchasing their own equipment because their fire services were not providing or replacing PPE. This is in contrast to the

34% of South European firefighters. Out of the total number of UK firefighters, only 4.4% responded that they purchased their own equipment because their service did not provide appropriate PPE against a significant 62% of Greek firefighters. The second highest number of firefighters stating that they purchased equipment because their fire service would provide or replace this was Portugal with 17.2%.

Firefighters were also asked whether they always wear the required personal protective equipment at every incident. For this question, the responses were relatively consistent between all countries with 'yes' responses ranging between 79% and 91%. In the sample groups, the variation of 'yes' responses between north and south European firefighters were 85.7% for north European firefighters (including UK firefighters) against 79% of south European firefighters. Within the south European sample group, Greece again had the lowest 'yes' responses at 66.5%, and the highest 'sometimes' responses at 25.2%. This was more than double the second highest 'sometimes' responses for a single fire service, which was the UK with 11.1% of 'sometimes' response.

In contrast with the above, when asked if fire officers at their fire station (crew managers / watch managers etc.) lead by example by always wearing their personal protective equipment at incidents, the variation in responses was more significant with 69% of north European firefighters (including UK firefighters) answering 'yes' against only 38% of south European firefighters. The most significant differences were again between UK firefighters and Greek firefighters with 86.7% of the former answering 'yes' against only 27.7% of the latter. The remainder of south European fire services were fairly consistent with 'yes' responses ranging between 44% and 50%. The sample group of north European fire services (excluding the UK) was also fairly consistent with the remaining south European fire services (excluding Greece) providing 43.7% of 'yes' responses.

When asked whether firefighters feel that there is a strong culture around safety in their fire services, the responses between the sample groups of north European firefighters (including UK firefighters) and south European firefighters were notable with 83% of north European firefighters and only 39% of south European firefighters attesting to having a strong safety culture in their fire service. The most outstanding differences were again between Greek firefighters (23%) and UK firefighters (87%). This can also be evidenced against documented standards around safety which are not so prominent is south European fire services.

Training

The extent and frequency of training forms the fundamental basis for firefighters to be able to undertake their work effectively and safely. It also helps firefighters keep pace with evolving technology and new challenges that arise as a result of modernisation and climate change. This thematic area aims at identifying firefighters' perceptions around the adequacy of training offered by their fire services in preparing them for their operational duties. In general, north European and UK firefighters provided higher numbers of positive responses that south European firefighters. These figures ranged between 61% for the sample group of north European (including UK firefighters) against 39% of south European firefighters. Out of the 61% of all north European firefighters (including UK firefighters), the UK firefighters as a single group provided the majority (66.7%) of positive responses. Greek firefighters provided by far the largest number of negative responses (i.e., they felt that their training had not prepared them adequately) at 71% and conversely the lowest number of positive responses at 29%. 49% of both Italian and Portuguese firefighters felt that their training had adequately prepared them. Spanish firefighters provided slightly lower positive responses at 40%. The sample group of north European firefighters (excluding UK firefighters) was slightly higher at 53%. Interestingly, Italian firefighters provided the highest number (31.4%) of responses under the category 'other'. Some of these responses did not provide further insight into Italian firefighters' perception around the adequacy of training their fire service offers them. However, a recurring theme was that training was either not realistic enough to simulate incident conditions or not adequate in preparing them for real incidents. The negative responses by Italian firefighters (i.e., their training was not adequate) were the lowest at 20% and on a par with UK firefighters. Spain and Portugal provided fairly similar negative response at 51% and 46%, respectively.

It appears that north European fire services (including UK fire services) are far more consistent at providing their firefighters with opportunities to further develop their training than are south European fire services. Over 80% of all north European firefighters (including UK firefighters) believed that their fire service does offer them opportunities to further develop their training as opposed to only 46% of all south European firefighters. The highest satisfaction rates were among UK firefighters (82%), although responses from other north European firefighters were similar at 78%. Greek firefighters provided the lowest numbers of positive responses with only 30% feeling that they were offered opportunities to further develop their training. Italian and Spanish firefighters also provided relatively low positive responses which ranged between 46% and 49% respectively. Portuguese firefighters provided higher levels of confidence in their fire services at 62.4%.

Based on the above responses, the next question was whether firefighters had undertaken training at their own expense since joining the fire service in order to compensate for inadequacies in training. The responses between the sample group north European firefighters (including UK firefighters) and the sample group south European firefighters were very similar, with 49% of the total number of north European firefighters (including UK firefighters) and 47% of the total number of south European firefighters stating that they had paid for further training. Out of the total number of north European firefighters, 78% were from the sample group north European firefighters (excluding UK firefighters), whereas only 28.9% of UK firefighters attested to paying for further training. Out of south European firefighters, Spanish firefighters were by far most prominently represented with 87.5% stating that they had paid for further training. This is proportionately the highest number of 'yes' responses of all firefighter groups. Other south European firefighter groups were relatively similar with 47% of Greek firefighters stating that they had paid for additional training against 34% of Italian firefighters and 35% of Portuguese firefighters.

In order to identify the reasons why firefighters undertook additional training at their own expense, five phrases were prepared from which respondents could select only one. These responses were:

- This doesn't apply to me.
- It further develops the skills I need for the job.
- I do not have this opportunity through my fire service.
- Other.

An error that occurred in the data collection process was that this question on the questionnaire was accidentally not locked to restrict responses to only one option. As a result, a few participants provided more than one response although the number of occurrences were relatively low. Out of the total number of firefighter respondents, 254 (44.1%) stated that the undertaking of training at their own expense does not apply to them, whereas 211 (37%) attested that the reasons they undertook additional training at their own expense was to further develop skills needed for their job. Only 63 (11%) of firefighters identified that the reason they undertook training at their own expense was because they do not have this opportunity through their fire service. This creates somewhat of a paradox because it would normally be expected that if firefighters were paying for additional training to further develop their skills required for the job, then this would mean that their fire service is either not offering them this opportunity or that the level, structure and content of the training offered by the fire service is objectively not fit for purpose. It could also mean that personal perception of firefighters undertaking further training, or the level of training required to specialise in certain aspects of their work (e.g., rope rescue or rapid water rescue), is different that the level of training currently offered by their fire service. These points, however, are not reflected through the responses.

It is acknowledged that a number of industry professionals that are highly regarded in the firefighting world offer private training programmes that on an intellectual and practical level fall outside the scope of process-driven training that would be typical of a fire academy. These training programmes provide additional science and evidence based insight that, to an ambitious and conscious firefighter or officer, will further enhance their competencies by giving them additional tools to better undertake their operational duties. In northern Europe (excluding UK firefighters), firefighters appear more inclined to pay for such training programmes to further develop their skillset. Comparatively, UK firefighters provided the majority of 'this doesn't apply to me' responses and the fewest 'it further develops the skills I need for the job' responses. This correlates with the previous question that demonstrated that the UK firefighters were the least likely to undertaken additional training at their own expense and most satisfied that their training adequately prepares them for dealing with the incidents they attend. Additionally, National Operational

Guidelines form a basis for the development and structure of relevant operational training programmes. UK fire services generally have a very structured approach to training and procedures for the maintenance of competence.

Regardless of whether these training programmes are objectively fit for purpose, they nonetheless do not allow much space for deviation and possibly do not instil a need for firefighters to search for additional training elsewhere.

However, having undertaken certified training is numerous countries, it is the author's opinion that the Compartment Fire Behaviour Training Instructors programme as certified by a third party certificating body and delivered by numerous training providers and fire services across the UK was, in 2019, missing important aspects of information relevant to fire behaviour and operational tactics that were conversely covered in other north European training courses. Nonetheless, the impact of this absence of knowledge cannot effectively be assessed and there is little evidence to date of near-miss incidents or injuries occurring in UK fire services as a result of this knowledge gap. It is also likely that dangerous events are not identified or recorded due to the lack of perception around the risks this absence of knowledge presents. In the case of south Europe, Greek and Spanish firefighters showed the highest number of respondents who attested that they undertook training at their own expense in order to develop the skills they needed for the job. These were 34.4% of Greek firefighter respondents and 65.3% of Spanish respondents, respectively. Not one Spanish firefighter responded, 'this does not apply to me'.

Due to an error in the questionnaire process, a number of firefighters included comments in the free text box reiterating the responses 'it further develops the skills I need for this job' and 'I do not have this opportunity through my fire service'. Collectively these additional responses, represent 14% of

Spanish firefighter responses whilst an additional 18% checked the box 'I do not have this opportunity through my fire service'. Strangely, 43.7% of Greek firefighters attested that 'this does not apply to me' and only 10.7% that 'I do not have this opportunity through my fire service'. It is widely acknowledged that the training offered in the Hellenic Fire Service is sub-standard to that offered in other European countries. This can be confirmed by the absence of suitable and modern training facilities whereby firefighters can adequately train for the incidents they are likely to attend during their service. A new national training facility was developed in East Attica through a private funding initiative and delivered in March 2023.

However, there is scepticism around the appropriateness of training props, the curriculum and whether these are adequate to prepare firefighters for the variety of incidents they will likely attend during their career. Through discussions with trainee firefighters, they mentioned that during their two-year study at the fire service academy, practical training was generally restricted to their internships at fire stations on weekends and this was often dependent on the shift and its appetite to deliver training. The result of this situation was that on-station training did not follow a specific curriculum and varied between stations and between shifts with some trainees receiving 'some training' and others receiving none at all. Furthermore, although it was mentioned by numerous trainees that they had not even bowled out hose during their first year at the training academy, they were given lessons on structural engineering. Specialist units such as the USAR team are given specialist training which is refreshed in accordance with the certification period.

Such courses include rope rescue, rapid water rescue, dog handler training, rescue diving etc. Unfortunately, such provisions have not readily been offered to response firefighters at urban fire stations although there has been some progress to address these matters. As such, the low responses relating to the training opportunities offered by the Hellenic Fire Service as well as 'This doesn't apply to me' responses appear somewhat paradoxical, as there is evidence that neither the level of training nor the opportunity to develop training is appropriate. To some extent, it could be surmised that the high numbers of 'this doesn't apply to me' responses from Greek firefighters could be explained as a result of either ignorance, or complacency, around the appropriate level of training required to undertake their job safely and effectively.

The following two questions were in reference to whether training provided by their fire service as well as the training for rank progression meet nationally recognised standards. The highest number of positive responses came from the group of North European firefighters (excluding UK) out of which 87.5% of firefighters agreed that their training meets nationally recognised standards. Similar scoring was achieved between the UK (73%), Italian (74.2%) and Portuguese firefighters (76.3%). As anticipated, only 48% of Greek firefighters and 36.1% of Spanish firefighters stated that firefighter training as provided by their fire services met nationally recognised standards.

It appears that where respondents have not confirmed whether their training meets nationally recognised standards, the remaining 'no' responses between most fire services account for the lowest percentage of responses, whereas it appears that a number of firefighter respondents are not aware of whether their training meets nationally recognised standards. The highest percentages of 'I'm not sure' responses were given by 35.9% of Greek and 40.3% Spanish firefighters. Conversely, the age group most prominently represented by 'I'm not sure' responses were again Greek (16%) and Spanish (16.7%) firefighters with 10 to 19 years of service. It is not clear why this is, and further research would be required to determine why firefighters are not aware of the training standards offered by their fire services.

Interestingly, the responses in respect of whether training for rank progression met nationally recognised standards were notably lower across all groups of respondents. 'Yes' responses were consistently the highest for North European firefighters (excluding the UK firefighters) and the highest number of 'no' and 'I'm not sure' responses were again given by Greek and Spanish firefighters. These percentages are generally consistent, with the idea that north European firefighters show the highest levels of confidence in their training standards whilst Greek and Spanish firefighters exhibit the lowest levels of confidence. The lower levels of confidence across the board in respect of training related to rank progression are also unclear and require further investigation. Some assumptions can me made around the level of confidence firefighters have in their commanding officers, transitions in training or higher levels of dissatisfaction in the training provided for rank progression by developing firefighters and officers. Further insight and other correlating factors might be identified later in the data analysis process.

The final question in this section was regarding the frequency at which training is undertaken in their respective fire services. Continuous professional development in the fire and rescue service is not only essential for firefighters to keep pace with changes in technology that ultimately have an impact on risk and the specific techniques that will be required to manage such incidents more effectively, but it is also essential that firefighters maintain competence in order to ensure operational preparedness and efficiency. The respondents were given a number of options to choose from. The frequency ranges were 'Rarely', 'Weekly', 'Monthly' and yearly'. Respondents were able to select more than option although it is acknowledged that the periodicity of some training regimes might fall outside these categories, such as daily, quarterly or bi-annually. Respondents were given a free text box under an option titled 'other' to indicate an alternative frequency although only 4% of firefighters from the sample group south-European firefighters chose this option, and this number represented 10% of Portuguese firefighters only. The two training frequencies that provided the most interesting data were the categories 'weekly' and 'rarely' whereas the option 'monthly' was broadly consistent for all sample countries. For the option 'weekly', 75.6% of UK fire fighters attested to undertaking weekly training, which was the highest number of responses among all respondents. This was followed by 57% of Spanish firefighters and 53% of firefighters in the north-European (excluding UK firefighters) sample group respectively. 40% of Italian firefighter respondents attested to undertaking weekly training, and Greek and Portuguese firefighters provided the lowest number of responses at 32% and 34%. The highest number of responses under the category indicating that they rarely undertake training was given by Greek firefighters at 28%.

The figures indicated under the option 'rarely' for the sample group of 'north European firefighters' (including UK firefighters) is statistically unimportant with only 1.3% of firefighters in this sample group selecting this option. The highest frequency of training across both sample groups appears to be weekly training regimes with 66% of firefighters in the north European sample group (including UK firefighters) selecting this option and 37% of firefighters in the south European sample group. These responses accounted for 40.8% (235) of all firefighter responses. From these data, it is apparent that north European firefighters are more likely to undertake weekly training that firefighters in southern Europe. Comparatively, firefighters in southern Europe (21.4%) were more likely to state that they rarely undertake training against only 1.3% of north European firefighters. The responses for this option represented 18.8% (108) of all firefighters responses. Lastly, for the option 'monthly', the responses were broadly consistent across both sample groups with 20.8% of north-European firefighters (including UK firefighters) stating that they take monthly training and 25.5% of south European firefighters. These response accounted for 24.8% (143) of the total number of firefighter responses.

Communications

There is critical information to be drawn from the data in this section. There is general consensus across both sample group categories for north (90%) and south European firefighters (93%) that radio communications at an incident are very important. There is also general consensus that personal hand-held radios for firefighters are very important with 80.5% of north European firefighters (including UK firefighters) and 78% of south European firefighters attesting to this. Data results become interesting on the forthcoming questions. When asked whether their fire service has issued them with personal hand-held radios, 67.5% of firefighters in the group 'north European firefighters' (including UK firefighters) indicate that this is the case.

UK firefighters were most likely to be issued with a personal hand-held radio where 73% of respondents state that they have been issued one. As there is no longer a national fire service standard in the UK, it is likely that not all fire services issue their staff with personal radios. Conversely, only 26.5% of firefighters in the sample group 'south European firefighters' attested to having been issued a personal hand-held radio. Interestingly, from this sample group, Spanish firefighters were an outlier with 87.5% having attested to have been issued with a hand-held radio. The remaining responses were not encouraging with 20.4% of Greek firefighter respondents, 17.1% of Italian firefighters and only 11.3% of Portuguese firefighter respondents attesting to having been issued a personal hand-held radio.

When asked if firefighters have ever purchased their own hand-held radio, not one firefighter in the sample group 'north

European firefighters' (including UK firefighters) claimed to have purchased their own hand-held radio. Contrarily, approximately 46% of firefighters in the sample group 'south European firefighters' claimed to have purchased their own hand-held radios with Greek firefighter respondents being most highly represented at 73% followed by Portuguese firefighters at 34.4%, Italian firefighters at 28.6%. Only, a very limited number (5.6%) of Spanish firefighters claim to have purchased their own hand-held radio.

The main reasons selected to explain why firefighters purchase their own hand-held radios are because they understand the important role this plays in their safety and because their fire service will not or has not issued them with one. It is clear and broadly accepted that radio communications provide a lifeline for firefighters when they are in distress and serve to provide incident critical information to and from the incident ground with the purpose of ensuring a more effective, efficient and safe response. However, similarly to other firefighter equipment (tools and the like), radios must be purchased through a central procurement process to not only ensure interoperability and consistency, but these should also satisfy specific technical specifications required for fire and rescue service operations (Killen et al, 2005). These important aspects cannot be satisfied if firefighters are purchasing their own hand-held radios. It will undoubtedly be the case that consistency will vary significantly, and it is more than likely that firefighters will resort to buying lower quality products that would be within an off-the shelf price range. Professional hand-held radios are generally over 4 - 5 times the price of an amateur hand-held radio. It is understandable that the quality and technical specifications will also not be equivalent.

In this section, the highest levels of dissatisfaction were present among Greek firefighter respondents. Greek firefighters were least likely to be provided with a personal hand-held radio by their fire service, they were most likely to have purchased their own hand-held radio when understanding the important role these play in their safety at operational incidents. Greek firefighters were also by far most dissatisfied with the adequacy of radio communications in their fire service with 71.3% of all Greek fire fighter respondents attesting this, with the second highest number of respondents stating that the quality of radio communications in their fire service was 'poor' being UK firefighters at 22%.

This notably high response from Greek firefighters would indicate that the standard of radio communications has unlikely to have undergone any particular improvements following Matthaios Mantzios fatality in 2014. It is also interesting that there does not appear to be consensus between the two group samples. It is apparent that, on average, north European firefighters (including UK firefighters) are more likely to be provided with a personal hand-held radio than south European firefighters with the exception of Spanish firefighters who are more likely to be provided with a personal hand-held radio, whereas south European firefighters are most likely to purchase a radio on their own in order to ensure their safety at operational incidents.

This is particularly pertinent to interior attack crews working inside smoke logged structures where there is a higher risk of disorientation and entrapment, where securing a point of contact with incident commanders and firefighters on the outside is critical (Mora, 2016). Different fire services utilise alterative configurations for radio communications. In a team of two firefighters, only one might be issued with a radio or a crew commander monitoring a team of two (or more) firefighters might be solely responsible for radio communications. There are studies discussed in this book under Chapter 2 that evidence the importance of robust communications in firefighter safety. It would be the organisation's responsibility to demonstrate how their radio communication procedures satisfy health and safety standards for firefighter protection. Conversely, with such evidence around the importance radio communications play in firefighter safety, it would be difficult to justify the absence of the provision of radios for frontline firefighters.

Incident Command

This section aimed at identifying the extent at which a defined command and control structure is present within each sample group fire service. Normally a command and control structure, in order to be effective, will need to be set against and work towards a range of standardised guidelines. Such guidelines or procedures will start at tactical level creating a structured (within reason) approach at first responder level, all the way up the chain of command to strategic level. This thesis is primarily concerned with command and control, standard operating guidelines, and procedures at tactical level as these will be critical to the progress and outcome of any incident from the offset. In other words, the outcome of an incident will depend heavily on how the incident is dealt with in the first minutes upon arrival. The less structured the approach, the higher the risk is of something going wrong (Campbell and Hall, 2022).

The first question in this section of the questionnaire was *"Does your fire service have standard operating procedures/guidelines for dealing with different types of incidents?"*. The variation of responses was not significantly different although 'yes' responses were more prominent for the group north-European fire services (including UK firefighters) at 88% against 61% for the sample group 'south European firefighters'. The UK firefighter (98%) and Italian firefighter (86%) respondents provided the highest 'yes' responses, whereas the Greek (59%) and Spanish (57%) firefighters provided the lowest number of 'yes' responses. However, all south European firefighter groups apart from Italy provided the highest number of 'No' and 'Some' responses.

When asked if incident commanders ensure the implementation of procedures and guidelines at every incident, the bulk of responses ranged between 'always' and 'sometimes', whilst the responses for the option 'never' were consistently the lowest for all sample groups. For the sample group 'north European firefighters' (including UK firefighters) 44% of respondents stated that incident commanders always ensure procedures and guidelines are implemented at every incident. Within this group, 62.5% of UK firefighter respondents chose the 'always' option whereas only 20% of firefighters in the group north European firefighters (excluding UK firefighters) believe that incident commanders always ensure the implementation of guidelines and procedures with 77% selecting the option 'sometimes'. This group and the sample group of Italian firefighters (90%) were most highly represented in the option 'sometimes'. Around 31% of respondents in the group south-European firefighters believe that incident commanders ensure that procedures and guidelines are implemented at every incident against 62% who believe that this only occurs sometimes. From the group of south European firefighters, the number of respondents selecting the option 'sometimes' ranged from 51% for Portuguese firefighters, 61% for Spanish firefighters and 69% for Greek firefighters.

What appears interesting, is that despite Italian firefighter respondents (86%) being most highly represented of all south-European sample countries as possessing procedures and guidelines for incidents they attend, conversely, they are most likely (90%) out of all sample groups to believe that incident commanders only sometimes ensure that procedures and guidelines are implemented. Even for the sample groups of the north European firefighters, although 98% of UK firefighters attest to having standard operating procedures and guidelines,

62.5% state that incident commanders will always ensure that these are implemented at every incidents whereas 37.7% believe that this is only 'sometimes' achieved. Similarly, although 75% of firefighters in the group 'north European firefighters' (excluding UK firefighters) attest to having procedures and guidelines, only 20% believe that their implementation is always ensured by incident commanders whereas 77% believe that this is only sometimes ensured.

When asked if respondents have a clear understanding of the Standard Operating Procedures and Guidelines for the different types of incidents they attend, the results were as anticipated. Around 76 % North European firefighters (including UK firefighters) were more likely than south European firefighters (52% of respondents) to provide positive 'yes' responses. The 'no' responses were generally low across all sample groups whereas the responses under the category 'some' were relatively high across most sample groups. UK and Spanish firefighters had the highest levels of 'yes' responses (85% and 77% respectively) and the lowest levels of 'some' responses - 12.5% for UK firefighters and 14% for Spanish firefighters. It could be presumed that these sample groups have either a higher level of training and familiarisation with their operational procedures and guidelines or that these are more strictly enforced, or potentially a combination of both. For the sample countries in south Europe (excluding Spain) as well as for the sample group 'north European firefighters' (excluding UK firefighters), it could be a cause for concern that firefighters do not have a clear understanding of SOPs/SOGs for different types of incidents seeing as these are designed to assist in more effective, efficient, and safe tactics. Assessing the effectiveness of samples groups that have indicated higher levels of uncertainty around procedures and guidelines is not easily measurable and would require further investigation at organisational level where such issues are identified in order to determine this correlation.

The next series of questions attempted to identify confidence levels felt by firefighting staff in their incident commanders. Interestingly, in these following questions, there appeared to be a general level of consensus between all sample groups of firefighters. The first question was whether respondents had faith in the competence of their incidents commanders. 39% of respondents in the sample group 'north-European firefighters' (including UK firefighters) attested to having faith in the competence of their incident commanders against 40% of south European firefighters. There was a predominance in the 'some' category of responses indicating that firefighters, irrespective of sample group, generally have faith in the competence of some incident commanders (rather than all) with 56% of firefighter respondents in the sample group 'north European firefighters' (including UK firefighters) and 51% of firefighters in the sample group 'south European firefighters' attesting this.

The next question was whether firefighter respondents have faith in the tactical decisions made by incident commanders. The responses for this question are almost identical to the previous question. There is some redundancy in this question due to a correlation between the mechanisms governing the responses. The lack of faith in incident commanders will almost certainly have to do with their decision making capabilities, although other factors relating to the intrinsic qualities and skills of the individual will also be pertinent.

Dynamic risk assessments are key to incident management. These are described as the assessment of risk in a rapidly changing environment at an incident where decisions are sometimes made in fast-moving situations, with incomplete or inaccurate information. It is a process not a control measure. The outcome of the assessemnt will contribute to the incident plan. It helps to inform whether personnel should be operating in the hazard area, which in turn determines the tactical mode (Fire Service Manual, 2008: p - 21). This process aims to ensure firefighter safety insofar as is reasonably practicable by limiting the extent of risk presented to firefighters at emergency incidents. Although all firefighters will undertake some form of dynamic risk assessment when operating at emergency incidents, the overall reponsibility for this task usually sits with the incident commander (London Fire Brigade, 2020) who will likely have a more holistic view and will receive and process multiple sources of infromation pertinent to the operation.

In the first of two related questions, respondents were asked to express their opinion regarding the extent of which they believe that the dynamic risk assessment process by the incident commander informs tactical decisions at an incident. The responses of the sample groups of both south European firefighters and north European firefighters (including UK firefighters) that believed that the dynamic risk assessment process by incident commanders 'greatly' informed tactical decisions at an incident were almost identical, with 70% of firefighters from the sample group 'south European firefighters' and 69% of firefighters from the sample group 'north European firefighters' selecting this option. However, for the sample group 'north European firefighters', UK firefighters (89%) were disproprortonately more likely than the sample group 'north european firefighters' (excluding UK firefighters) (40%) to select the option 'greatly', whereas the sample group 'north european firefighters' (excluding UK firefighters) were more likley to select the option 'moderately' (53%) and were in fact the sample group most highly represented in the option 'moderately'. Firefighters from Greece (81%) and Italy (83%) were also highly represented in the option 'greatly', indicating that, in their opinion, dynamic risk assessments undertaken by incident commanders greatly informed tactical decisions. Portuguese firefighters followed with 66% of firefighter respondents selecting the option 'greatly'. Spanish firefighters were on par with the sample group 'north European firefighters' (excluding

UK firefighters) with 46% slecting the option 'greatly' and 46% selecting the option 'moderately'.

Out of all sample groups of firefighters, 70% of respondents believed that the dynamic risk assessment process undertaken by incident commanders greatly informs tactical decisions at incidents whilst 27% of firefighters from all sample groups believe that that this occurs in moderation. A total of 3.5% of all firefighter respondents selected the option 'Not at all'. However, this figure does have statistical meaning and would likely be an exception based on personal experience of the particular repondents rather than an organisational trait.

When asked whether tactical decisions made by incident commanders prioritise firefighter safety there was slightly higher confidence (68%) in the sample group of 'north European firefighters' (including UK firerifghters) than for the sample group of south European firefighters (52%). The highest confidence levels were among Italian firefighter respopndents with 86% feeling that tactical decisions by incident commanders prioritse their safety, followed by 71% of UK firefighter respondents. The lowest confidence levels were among Greek firefighter respondents with only 34% feeling that incident commanders prioritise their safety when making tactical decisions. Unsurprisingly, Greek firefighter respondents were also most likely to attest that incident commanders do not always prioritse their safety.

Firefighter respondents from Spain, Portugal and the sample group north European firefighters (excluiding UK firefighters) provided relatively consitent responses for both 'Yes' and 'Not Always' responses with a slightly higher inclination towards 'Yes' responses. Nonetheless, there is value in understanding why a notable number of firefighters feel that their safety is not always considered in tactical decisions of incident commanders. Further research would be required to fully understand the intrinsic aspects leading to this belief. However, it is assumed that tactical decisions that have put firefighters at risk could explain why they only have faith in the competence and tactical dercisions of some incident commanders.

It is broadly accepted that the firefighter's occupation is inherently dangerous and is presented with health and safety risks at every incident (Hocy, 2023; USFA, 1993). The previous question explored firefighter respondents opinions as to whether incident commanders prioritise firefighter safety when making tactical decisions. It could be argued that the results gleaned from the respondents were not overly encouraging as the proportion of all firefighters agreeing that this is the case (54%) against those that expressed that this is 'not always' the case (39%) were quite similar when considering both the legal and ethical importance of ensuring firefighter safety. Assuming that firefighter safety is fundamentally considered by incident commanders, the next question aimed at identifying whether the organisations of the firefighter respondents have a documented procedures for dealing with firefighter emergencies. The sample group 'north European firefighters' (including UK firefighters) scored disproportionately higher than the group 'south European firefighters' with 77% of the former identifying that they have documented procedures for firefighter emergencies against only 28% of the latter sample group. UK firefighter respondents provided the highest number of 'yes' responses (89%) followed by the Italian firefighters (74%). Although fire services in the UK historically possessed documented procedures for firefighter safety, this became even more accentuated following serious incidents leading to firefighter fatalities such as the King's Cross fire in London (1987), the Shirley Towers fire in Southampton (2010) and other similar incidents.

Roughly 59% of north European firefighters (excluding UK firefighters) state to have documented procedures for firefighter emergencies, conversely, 34% of firefighter respondents from this state that they do not have any such procedures. Greek firefighters provided the lowest number of 'yes' responses with only 19% stating that they have documented procedures for firefighter emergencies whilst 29% state that they do not and 52% that they are not sure. In the first instance, these figures indicate that if there are documented procedures for firefighter emergencies, with such low confidence levels, they are unlikely to be known, trained against, and implemented in practice in the event of a firefighter emergency.

This was blatantly apparent in the Mantzios Case Study on which this book is structured, and research undertaken by the author has not identified any such documented process by the Hellenic Fire Service. Portuguese firefighters also scored low numbers with only 28% indicating that they possess documented procedures for firefighter emergencies. 44% stated that they do not have any such procedure and 28% were not sure. Likewise, 32% of Spanish firefighters identified that they have documented procedures for firefighter emergencies whilst 50% stated that they do not have any and 18% of respondents stated that they were not sure. Apart from the figures for UK and Italian firefighter respondents, the figures for the remaining sample countries are not encouraging. It can be reasonably surmised that documented procedures for firefighter emergencies are not a de-facto policy in all fire service organisations and in cases where they might be, the relatively high 'I'm not sure' responses would indicate that these are not communicated to all staff and their implementation in an emergency situation will unlikely occur – at least in any structured format.

The last question in this section provided two phrases of which respondents had to choose the one that best describes them. These phrases were as follows:

a. *"At Incidents, I work in accordance with the standard operating procedures / guidelines dictated by operational policy both of which are reflected through my training."*

b. *"At incidents, I work to the best of my capabilities with the knowledge, training and experience I have gained as a firefighter, but not necessarily in accordance with standard operating procedures / guidelines".*

The most statistically significant data was acquired by the sample group of Greek firefighters, of which 98% attested to working to the best of their capabilities with the knowledge, training and experience they have gained as a firefighter, but not necessarily in accordance with standard operating procedures / guidelines. Despite 59% of Greek firefighters identifying that their fire service does have standard operating procedures or guidelines for different types of incidents, it is apparent that these do not inform how firefighters will operate at an incident. This statistic comes as no surprise, as this research has not identified the implementation of any documented procedures, at least at run of the mill incidents. In addition, the only operational protocols that have been identified are covered under Fire Service Circular 116 where a hand-full of these have been documented although it is unlikely that these are fit for purpose or implemented on a practical level. Conversely, although 98% UK firefighters identified that their fire service has standard operating procedures or guidelines for different types of incidents, 58% of UK firefighter respondents selected option (a) whilst 38% chose option (b). Although UK firefighter respondents were the most highly represented in option (a) indicating a direct correlation between tactics and procedures, the number of firefighters selecting option (b) is not negligible and might indicate a greater affiliation to learned processes and actions through experience and repetition (Gough, B. cited in Grimwood, 2017: p - 443).

This equally applies to Italian firefighter respondents. Although 86% identified that their fire services possess standard operating procedures or guidelines for different types of incidents, 74% selected option (b) rather than option (a) as would have been anticipated. Firefighters in the sample group 'north European firefighters' (excluding UK firefighters) were also highly represented, with 78% of respondents selecting option (b). Overall, figures for option (b) were notably higher, with 67% of firefighters in all sample groups choosing this option over the 28% of those choosing option (a). Based on these figures, there appears to be an inclination for firefighters to rely on skills gained through experience rather than through the implementation of procedures and guidelines. This does not mean that firefighters do not implement procedures at all, but perhaps that aspects of these are chosen and applied where this is felt appropriate. It could also indicate that procedures available to firefighters are not directly applicable to incident demands, therefore requiring alternative approaches that are deemed effective and then recycled.

Firefighter Health, Safety and Well-being

The first question in this section of the questionnaire aimed to determine firefighters opinions on their salaries in an attempt to identify a correlation between performance, job satisfaction and remuneration. The relationship between these three elements was considered fundamental in how firefighters assess the relationship between their salary and job related risks (i.e., are they paid enough for the risks they take). However, research into this area identified that, in general, economic incentives are not found to be the main motivators for job performance and in some cases, in fact, they are even counterproductive (Pugno and Depedri, 2009). Other incentives such as promotion (Rinny, Purba and Handiman, 2020), a genuine interest in the area of employment, a positive work environment (Rachman, 2021) and good leadership (Belias and Koustelios, 2014) appear to be more relevant. In this regard, although there are several factors that will collectively influence job satisfaction, the correlation between job satisfaction and performance appears to be clearer although, in itself, this relationship is not straight forward. For these reasons, firefighter remuneration will not be considered as part of this book.

When asked if fire services take firefighters personal and work needs into consideration the majority of firefighters (43%) neither agreed or disagreed, implying that perhaps some needs were considered and some not. Italian firefighters provided more definitive responses with 60% disagreeing and 40% agreeing that their fire service considers their personal and work needs. These two statistics were the highest among all sample groups, although it should be kept in mind that due to the small number of responses, these figures should be reviewed with caution. Responses in the 'Neither Agree or Disagree' option were generally similar amongst all sample groups (except Italian firefighters at 0%) that ranged between 40% and 46%. Greek firefighters provided the highest number of responses for this option.

When asked if fire services take firefighters health, safety and well-being into consideration, the collective majority (40%) agreed, 33% of all respondents neither agreed nor disagreed and 27% disagreed. The highest numbers of firefighters that disagreed that their fire services take their health, safety and well-being into consideration were among the sample group south European firefighters at 30% against only 10% of north European firefighter respondents (including UK firefighters). Respectively, the sample group north European firefighters (including UK firefighters) were most likely to agree that their fire services take firefighter health, safety and well-being into consideration (63%). The sample group north European firefighters (excluding UK firefighters provided the highest number of 'Agree' responses (72%) followed by the sample group UK firefighters at 58%.

The numbers of south European firefighters agreeing that their fire service takes fire fighter health, safety and well-being into consideration were not significantly different that the responses for UK firefighters with 54% of Italian firefighters, 49% of Portuguese firefighters and 40% of Spanish firefighters. Greek firefighters provided the lowest number of 'Agree' responses with only 20% believing that their fire service considers firefighter health, safety and well-being. Greek firefighters were also most highly represented in the option that neither agreed or disagreed that their fire service takes their health, safety and well-being into consideration. Italian firefighters (46%) were most likely to disagree that that their fire service takes firefighter health, safety and well-being into consideration followed by Greek firefighters (38%). Determining the intrinsic detail behind these responses requires further investigation as the extent and the aspects relating to the health and safety of firefighters are multi-faceted and include elements that extend beyond the research capacity of this study. Significant information can also be ascertained through a qualitative study. However, the next series of questions were aimed at providing further insight into firefighters perspectives on organisational health and safety.

Large-scale incidents are those that will most likely put fire and rescue personnel under significant physical and mental stress. In order to be effective as an organisation at large scale incidents, tactical goals can only be achieved through adequate management arrangements for staff health, safety and well-being. Throughout the next series of questions, there appears to be a distinctive contrast between north European firefighters (including UK firefighters) and south European firefighters with the former providing greater positive responses and lesser negative responses than south European firefighters. These questions aimed to identify whether there were adequate management arrangements in place at large scale incidents for regular rest for firefighters and staff relief, appropriate mechanical and human resources, the provision of food and drink, and medical support for on-scene firefighter injuries.

In the first instance, the majority of responses (42%) amongst all sample groups regarding regular rest for firefighters and staff relief at large scale incidents were indicated as occurring 'sometimes'. These percentages were relatively consistent between Greek (45%), Portuguese (51%) and UK (42%) firefighters that represented the highest number of 'sometimes' responses against 28% of north European firefighters (excluding UK firefighters), 23% of Italian firefighters and 21% of Spanish firefighters that provided the lowest numbers of 'sometimes' responses that were also fairly consistent. Conversely, 38% of all sample groups indicated that regular rest and staff relief at large-scale incidents did occur with Greek firefighters (30%), Portuguese firefighters (27%) and UK firefighters (44%) providing the lowest number of positive responses and north-European (excluding UK firefighters), Spanish and Italian firefighter respondents providing the highest numbers of positive responses that ranged between 60% and 66%. Approximately 21% of all firefighters identified that there are not adequate measures at large incidents to ensure regular rest for firefighters and staff relief with the lowest number of 'no' responses amongst north European firefighters (including UK firefighters) and the highest number of 'no' responses amongst south European firefighters at 22%. Greek firefighters were most highly represented in this option with 25% of Greek firefighter respondents selecting the 'no' option to this question.

It would be important to determine whether these responses have been given against a comparison with a policy standard or workplace legislation, or whether these are based on personal opinion. If it is the former then it would be interesting to understand why, if there are policy standards that determine the provisions available to staff in relation to regular rest and relief at large scale incidents why these are not implemented in all such cases. Where such standards are not available there will inevitably be an element of subjectivity in the response provided. These could be intuitive in nature and be based on the reasonable amount of time that a firefighter can work at an incident without rest and without relief.

It is also acknowledged that underlying factors at major incidents might not allow for the normal operation of logistics and resource management. Jiang and Yuan (2018) discuss the achievements and challenges of emergency logistics in a large-scale disasters. It is acknowledged that, unlike routine emergency incidents, large-scale events might create significant challenges for emergency services to bring the incident under control, prioritise and provide adequate assistance, perform, rescues and so on. In this respect, resources might be stretched and staff over-overworked meaning that policies that would normally apply at standard emergencies might not be applicable at a large-sale incident due to increased demands in contrast with organisational capacity – at least until the incident or incident management process stabilises. Therefore, the responses to this question must be reviewed with some scepticism. The data itself does have merit, as large-scale incidents are proportionately fewer than routine emergencies where most fire services have a proportionate risk-based response model where regular firefighter rest and staff relief will have been considered and will take place. As large scale incidents will incorporate a number of dynamic factors and present a series of challenges for the emergency services, it is likely that the 'sometimes' responses to this question given by firefighter responses could be in respect of such scenarios like at large forest fires, industrial fires or accidents etc.

Another key factor relating to firefighter health, safety and well-being is the adequacy of mechanical (firefighting vehicles or specialist equipment) and human resources. If both of these resources are insufficient, there will be increased challenges for firefighters to bring a large incident under control. As per the previous question, collective responses were fairly similar with 44% of all firefighters stating that mechanical and human resources are adequate, 39% identifying that that these are sometime adequate and 17% attesting that these are not adequate. The lowest number of positive responses were from Greek (34%) and Portuguese (34%) firefighters. The highest proportion of 'yes' responses were given by 84% of north European firefighters (excluding UK firefighters). Greek (46%) and Portuguese (51%) firefighters provided the highest number of 'sometimes' responses that were significantly higher than any of the other firefighter sample groups. Greek firefighters (21%) also provided the highest number of negative responses, followed by UK firefighters (18%), Italian firefighters (17%), Portuguese (15%) and Spanish (13%) firefighters. The group of north European firefighters provided the lowest proportion of 'sometimes' (9%) and 'no' (6%) responses. This evidence is not statistically different than the previous question where the same correlation between resource management and challenges of large-scale incidents would also be pertinent.

When asked whether there are adequate management arrangements in place for the provision of food and drink at large incidents, the responses again did not provide any statistical variation from the previous questions. 44% of all firefighters agreed that there were adequate arrangements in place by their fire service, 40% selected the option 'sometimes' whereas 16% provided a negative response. North European firefighters provide the highest number of positive responses (75%) followed by Italian firefighters (60%). The remaining sample groups ranged between 40% - 47% except for Spanish firefighters that provided the lowest number of positive response at 32%. Although there was a distinctive difference between the sample group north European firefighters (29%) (including UK firefighters) and south European firefighters (42%), around 40% - 45% of Greek, Spanish and Portuguese firefighter respondents comprised these numbers whereas only 29% of Italian firefighters provided 'sometimes' responses. Spanish firefighters (25%) provided the highest number of negative responses followed by UK firefighters (20%).

With regard to the provision of medical support for firefighters at large incidents, although the majority or responses were positive, there are significant differences between fire services. 73% of the sample group of north European firefighters (including UK firefighters) attested to having medical support at large incidents against only 42% of south European firefighters. Italian firefighter respondents (71%) provided similar response proportions to firefighters in the sample group north European firefighters (including UK firefighters) whereas Greek (32%), Portuguese (49%) and Spanish firefighters (39%) provided the lowest numbers of positive responses. North European firefighters (19%) (excluding UK firefighters), UK firefighters (16%) and Italian firefighters (17%) provided the lowest numbers of 'sometimes responses' whereas 'sometimes' responses for Greek, Portuguese and Spanish firefighters were almost twice as high. Conversely, Greek firefighters (37%) provided the highest number of negative responses, followed by Spanish firefighters (25%). The low proportion of negative responses for other fire services can perhaps be explained as a result of the ambulance service being embedded within fire service as is the case in northern Europe and Portugal. Ambulances or medical support teams might also be automatically despatched to certain types of emergency incidents as part of a pre-determined attendance policy. Firefighters in Greece are neither provided with first aid training as part of their basic training, nor is an ambulance or medical team sent to large incidents by default unless so required or requested. In should be noted that the Hellenic Fire Service does have its own ambulance with adequately training staff.

Staffing numbers of fire appliances are also critical to firefighter health, safety and well-being. It has been determined earlier in this book that emergency incidents by type will also have critical tasks that must be achieved. These critical tasks by incident type will commonly determine adequate crew sizes and resources that are to be deployed to an emergency incident. Crew size is critical because if incident tasks require 5 or 6 staff to address safely and efficiently, a smaller crew size will mean more work for fewer people, increased response times and an elevated risks for both firefighters and the public. Out of all respondents, 54% of Italian, 53% of Spanish, 50% of Greek and 47% of UK firefighters identified that fire appliance staff numbers are inadequate. It is known that with the introduction of Integrated Risk Management Planning in the UK under the Fire and Rescue Services Act, 2004, staffing numbers were dramatically reduced on front-line fire appliances across many fire services in the UK (Cheshire Fire Brigade Union, 2017). This reduction was intended to reduce costs while maintaining an appropriate ability to respond (HMICFRS, 2018: 11). However, evidence put forward has demonstrated the risks that this presents (Grimwood, 2008). As there is no policy or standard for crew sizes in the Hellenic Fire Service it comes as no surprise that Greek firefighters provided the highest number (40%) of 'It may vary' responses. Apart from the sample group north-European firefighters (69%) (excluding UK firefighters) and Portuguese firefighters (52%), the low numbers of positive responses amongst all other sample groups (especially for Greek firefighters) are noteworthy.

The area of primary response is important for several reasons. Smaller response areas mean faster response times leading to more effective incident control. It also means that fire appliances will have to travel less time over smaller distances which reduces the risk for response crews. Inner city fire stations might have a higher risk of collision due to congestion but their overall travel speeds during peak times will be slower than fire stations located on the periphery of the city or in rural areas. The sample group north European firefighters (including UK firefighters) had by far the smallest primary response areas (61%) between 2.5km^2 and 12.5km^2 against only 9% of the sample group of south European firefighters.

Conversely, the sample group south European firefighters (39%) attested to having the largest primary response area at over 40km^2 against only 4% of the sample group north European firefighters (including UK firefighters). Greek (59%), Italian (69%) and Spanish (57%) firefighter respondents were among those that reported having the largest primary response areas at over 40km^2. These figures were the highest against all response area profiles. The highest proportion of Portuguese firefighter respondents (52%) attested to having a primary response area of between 25km^2 and 37.5km^2 whereas 31.5% reported having a response area of between 12.5km^2 and 25km^2. The comparison between north European (including UK firefighters) and south European firefighters were fairly consistent amongst all firefighter respondent groups with south European firefighters presenting higher number of responses for the response areas under the categories 12.5km^2 and 25km^2 and 25km^2 and 37.5km^2. Overall, it can be seen that South European firefighters are more likely to have larger primary response areas than north European and UK firefighters. This would mean longer response times and potentially fewer available resources in the early stages of an emergency incident.

Subject to the above findings, it would be assumed that firefighter responses regarding the adequacy of response times to emergency incidents within their response area would be reflective to the time of travel required within those areas. North European firefighters (64%) (including UK firefighters) were more likely to indicate that their response times were adequate than south European firefighters 38%. Out of the sample group of south European firefighters, Portuguese firefighters presented that highest number of 'adequate' responses which would be reflective of their slightly smaller response areas when compared to their other south European counterparts.

The results for the category indicating that response times may vary depending on the location of the incident were diametrically opposite with south European firefighters (58%) most likely to select this option against 32% of north European firefighters (including UK firefighters). The results attesting to slow response times were not as anticipated with statistically unimportant data achieved for this category. However, it is likely that where fire services do not have an average response time quota, then responses to this question would hold a higher level of subjectivity. It is known that north European and UK fire services have average response time quotas to emergency incidents – especially within urban areas. This does not appear to be the case, at least partially, in southern Europe. For example, there is no average response time quota in the Hellenic Fire Service. This raises a wider issue around effectiveness and efficiency. Fire services would normally need to determine, via statistical evidence based on both quantitative and qualitative criteria, what their average response times to emergency incidents should be in order to be in a position to measure their performance and respond within a timeframe that will not allow the incident to grow beyond the control of the resources that have been mobilised - insofar as is reasonably practicable.

In respect of response times for additional resources to emergency incidents, as anticipated, fire services that have smaller response areas and faster response times are more likely to receive additional resources within adequate timeframes. In this case, north European firefighters (64%) (including UIK firefighters) were more likely to attest that additional resources response times were adequate as opposed to 31% of south European firefighters. South European firefighters (58%) were notably more likely to select the option that additional resource response times to incidents will vary depending on the location of the incident than north European firefighters (26%) (including UK firefighters).

Out of the sample groups of south European firefighters, Greek firefighters (68%) were most highly represented in this option. UK firefighters were most likely to indicate that response times for additional resources were inadequate although further investigations into the raw data revealed a wide geographical distribution including city and county firefighter respondents. County fire services will have different response times than inner city fire stations. They are also more likely to implement a retained duty system with on-call firefighter meaning that overall response times will be slower than a staffed fire station that at the station and ready to go once the incident is despatched.

When asked if firefighters are under stress due to a lack of available resources when attending incidents, the responses were fairly consistent across the board. 30% of north European firefighters (including UK firefighters) and 39% of south European firefighters agreed that they are under stress due to a lack of available resources. From this category, 0% of Italian firefighters provided a response whereas Greek firefighters (48%) were most highly represented. 31% from the sample group north European firefighters (including UK firefighters) disagreed that they are under stress due to the lack of available resources, and this would be associated to the smaller response areas, the faster response times and the proximity of additional resources. However, the sample group UK firefighters provided responses to this category that were not dissimilar to other firefighter responses from south European firefighter sample groups.

Greek firefighters presented the lowest number of 'disagree' responses, whereas all other firefighter sample groups presented fairly proportionate responses despite only 15% of the sample group south European firefighters disagreeing to being under stress due to that lack of available resources. South European firefighters (46%) provided higher responses to the category 'this is sometimes the case' against 39% of north European firefighters (including UK firefighters). Italian firefighters were most highly represented in this category with 74% of firefighter respondents attesting that they are sometimes under stress due to the lack of resources. All other responses for the firefighter sample groups were fairly consistent.

Where line of duty deaths, injuries or near-miss events occur, there is an opportunity to analyse these and determine whether any lessons can be learned to not only educate front-line staff and policy makers but principally to ensure that every effort is made to avoid a similar events from occurring in the future. If this process does not occur, then firefighters might repeat similar errors unknowingly thus increasing the risk to their health and safety that could have otherwise been avoided. With regard to the Mantzios case study analysed in this book, there is no formal account of the mechanisms that led to Mantzios's death. This is neither apparent in the internal fire service investigation report nor in the technical report undertaken by an independent specialist on behalf of Mantzios's family.

Further near-miss events have occurred at similar incidents as a result of the same errors as those that occurred at the fire in 2014 that killed Matthaios Matzios. The single most important data arising from the question pertaining to whether there is a formal process by which lessons learned from serious incidents are made available to firefighters was that the sample group north European firefighters (including UK firefighters) provided disproportionately higher 'yes' responses (74%) that the sample group of south European firefighters. The highest proportion of positive responses were given by UK firefighters (84%). This was anticipated, as there have been detailed investigation reports for almost every Line of Duty Death Incident at least in the last 20 - 30 years which has in turn assisted in reforming operational standards to mitigate the same errors from re-occurring. The equivalent 'yes' responses were significantly lower for the sample group south European firefighters with only 20% identifying that there is a formal process by which lessons learned from serious incidents are made available to them. Greek (15.5%), Portuguese (18%) and Spanish firefighters (22%) provided the lowest number of 'yes' responses whereas Italian firefighters were significantly higher with 54% providing 'yes' responses.

When asked whether lessons learned from firefighter injuries or fatalities have assisted in reforming operational procedures to ensure that those same mistakes do not repeat themselves in their fire services, again, north European fire fighters provided greater 'yes' responses (73%) than the sample group south European firefighters (47%). This data is somewhat paradoxical as it has already been demonstrated that there are significantly low confidence levels among south European firefighters (excluding Italian firefighters) in identifying a formal process by which lessons learned from serious incidents are made available to them.

Although the statistics for the sample group north European firefighters are consistent with UK firefighters providing the highest number of 'yes' responses as per the previous question, conversely, 96% of Spanish firefighter respondents agree that lessons learned from firefighter injuries or fatalities have assisted in reforming operational procedures to ensure that those same mistakes do not repeat themselves although only 22% state that there is a formal process by which lessons learned from serious incidents are made available to firefighters.

Greek firefighters (29%) consistently provided the lowest proportion of 'yes' responses to this question, the highest number by far of 'I don't know' responses (40.7%) and the second highest number (30%) of 'no' responses. Italian firefighters (71%) consistently provided a high number of 'yes' responses in the group south European firefighters that would align with their responses in respect of having a formal process by which lessons learned from serious incidents are made available to firefighters. Apart from Italian firefighters, the differences between north European and South European firefighters insofar as possessing formal procedures for lessons learned following a line of duty death, injury, or near-miss incident and how these lessons are communicated to staff and assist in reforming operations standard is astounding.

Chapter 6 – Discussion

This chapter provides a comprehensive discussion of the findings derived from the comparative analysis of fire services across different European regions. The discussion explores the implications of responses received from firefighters in the UK, northern Europe, and southern Europe, focusing on key organisational, operational, and safety-related themes. The responses are analysed against a series of benchmark standards, revealing notable trends, discrepancies, and commonalities between the different regions.

Central to this discussion is an examination of how structural, cultural, and procedural differences influence the effectiveness and perceived reliability of fire services. Particular attention is given to the variations in standard operating procedures, training, leadership trust, and access to resources such as personal protective equipment and communication tools. The analysis also revisits the original hypotheses proposed at the outset of this study, testing their validity through the data obtained.

This chapter further aims to contextualise these findings within broader frameworks of organisational efficiency, leadership theory, and cultural dynamics. By doing so, it seeks to identify strengths, highlight shortcomings, and suggest areas for improvement, with particular emphasis on the contrast between UK and Hellenic fire services, which emerged as the most divergent in the dataset. Ultimately, this discussion lays the groundwork for the conclusions and recommendations presented in the final chapter.

The categories of questions that were selected represent aspects that are fundamental to the structure of fire service organisation, the efficiency and effectiveness of fire service operations as well as for the safety of fire and rescue personnel.

The responses to the questionnaire present clear differences between north European, UK and south European firefighters. In general, south European firefighter responses presented greater negative responses than north European firefighters (including UK firefighters) to questions where a 'No' response would carry a negative weighting (i.e. In your opinion has your training adequately prepared you for the incidents you attend?)

They were also more highly represented where 'yes' responses held a negative weighting (i.e. Have you ever purchased your own personal protective equipment?). Interestingly, there were some questions where responses were very similar across all sample groups. This was particularly the case for questions regarding perceptions around the competence of incident commanders and the extent to which firefighters have faith in their decisions where it appears that most firefighters trust some incident commanders and not others.

There were also some outliers in the group 'south European firefighters' where respondents from some countries represented unusually high 'yes' and 'no' responses where these had a positive weighting. For example, 87.5% of Spanish firefighters were most likely to be provided with personal hand-held radios and least likely to purchase their own.

These figures were the highest in the data set and even higher than the responses provided by the UK firefighter sample group. Italian firefighters were also most highly represented in the questions relating to the provision of standard operating procedures where almost 86% of respondents attested to having these. Such outliers were found to be atypical in contrast with other responses for the sample group of south European firefighters.

During the analysis of the data, it was noted that the responses to the questions could be categorised in the following ways:

- **Definitive responses** – These are responses that are objectively measurable, for example, does your fire service have standard operating procedures or has your fire service provided you with a hand-held radio where the answers are generally binary (Yes or No).

- **Fluid Responses** – These are responses that are ambiguous i.e. not definitive, where respondents have provided 'maybe' or 'I'm not sure' responses. These responses were somewhat paradoxical as where questions would normally require a definitive response i.e. either yes or no, like, for example, whether fire services possess standard operating procedures, ambiguous responses such as 'I'm not sure' would indicate a deeper issue that requires further investigation. This finding aligns with Choo (1996), who argues that organisations often suffer from "learning disabilities" when communication systems fail to structure and disseminate knowledge effectively. Additionally, Senge (1990) warns that many organisations are unable to become true "learning organisations" due to inherited assumptions and resistance to questioning past practices.

- **Perceptive responses** – these are responses that are based on personal beliefs, opinions or perceptions, such as whether firefighters have faith in the tactical decisions made by incident commanders.

Across all sample groups, UK firefighters provided the most consistent responses that, in most cases, were as anticipated subject to publicly available information regarding their structure, policies and processes that are regulated through national health and safety legislation and informed by National

Operational Guidance. As such, the sample UK firefighters provided the highest number of definitive responses and the lowest number of fluid responses. In general, fire services in the sample group 'south European firefighters' provided the highest number of definitive responses with a negative weighting – with only some exceptions. The responses provided from the group north European firefighters must be viewed with caution as responses were provided by firefighters from several north European countries.

Apart from this sample group being relatively small, it is likely that there are notable variations between fire services that might skew the data for this sample group although there were commonalities with UK fire services in some instances. The most apparent differences to the proportion of questions per category were between UK firefighter respondents and Greek firefighter respondents where, in many cases, these were counter-opposite.

Where questions required a definitive response it is interesting that fluid answers have been provided. One would expect that for a question regarding whether training meets nationally recognised standards or whether there are emergency procedures for firefighters, the answer with either be yes or no. The fact that there is uncertainty around these and other questions alike would indicate a broader communication issue i.e. fire services might possess information that is not communicated effectively at all staff levels.

Another interesting aspect identified in the data analysis is where definitive responses are 'error' responses and probably fall within the category of perceptive responses. For example, over 58% of Greek firefighters state that their fire service does have standard operating procedures or guidelines for dealing with different types of incidents when no evidence has been found in this research to demonstrate that this is true.

A more realistic definitive response would have been 'some' as there are a handful of procedures drafted under Fire Service Circular 116 but these do not cover volume incidents such as lift entrapments, structural fires, vehicle fires and so on. Such procedures and many others alike exist in many north European fire services. This reflects what Hartin (2013) describes as reliance on experience without understanding, a limitation endemic in many fire services that have not embraced evidence-based practice. Similarly, Sandlebeen (2002) and Denef et al. (2008) argue that firefighters often adopt habitual tactics learned on the job, regardless of whether these methods are the most effective or safe.

In Belgium for example, these procedures are even available to every firefighter via a mobile application although they are principally practiced in training drills and systematised as a result to be implemented at operational incidents. This is very much the case in other fire services reviewed as part of this study and in particular UK fire services. It was understood through interviews with Greek firefighters that drill type training against task-based procedures is not undertaken at the fire service college and that operational duties are implemented ad hoc.

The fundamental basis for procedures (or guidelines) is that it allows for the implementation of a process-driven methodology for dealing with different types on incidents where each team member broadly understands their role when arriving on scene. This is done for speed, efficiency and safety but it also allows for a level of action predictability (what each firefighter will do), accountability and to avoid freelancing practices which are, as discussed earlier, inherently dangerous. As Grimwood (2017) and Hartin (2013) argue, freelancing—where firefighters act independently due to unclear protocols or inadequate training—can result in serious safety risks. These practices are often justified through tradition rather than validated through structured learning.

Where perceptive responses have been given to these questions, these might also be as a result of a failure in communication or an absence of relevant knowledge. Although there are various high-level strategies for incident management in the Hellenic Fire Service, these are not designed as procedural documents as they do not include detail at firefighter task level. There are some training manuals, most notably those authored in 1994 and 1996 that broadly discuss processes for hose laying for example. These documents are not standard operating procedures but might be perceived as such in absence of anything else.

It could be argued that perceptive responses are influenced by the extent of knowledge held either individually or collectively within a defined group. Narens argues that, in the theory of belief, greater ambiguity produces less definiteness and vice versa (Narens, 2003). In addition, Lupyan (2016) purports that 'perception can only be reliable if it represents the world as it really is, a task that would be interfered with if what we see was influenced by expectations, knowledge, or beliefs' and, thus, 'what we perceive is strongly influenced by what we know'. In this respect, perceptive responses cannot be viewed as ultimately reliable and can be seen as contextual to each fire service. This is also pertinent to questions that required opinions on subject related matters, where those opinions demonstrated a level of variation between fire services.

In general, throughout the entire data set, there was a very limited number of responses that represented 100% of any single sample group's respondents. This was the case for one or two questions where all firefighters in the sample group 'north European firefighters' (including UK firefighters) were represented against one response. The level of ambiguity in the responses, although variable, could be attributed to several reasons. The first might be that firefighters, depending on their region might experience certain organizational and operational aspects differently. For example, a firefighter in the north of England might experience certain elements differently than a firefighter in London, despite the existence of National Operational Guidelines. This could also be directly linked to the differences in how fire services operate following the implementation of the Fire and Rescue Services Act, 2004. This could equally apply to any other fire service in the sample groups where variability in the way 'things are done' exists i.e. a firefighter at a metropolitan fire station in Athens might be more versed with a systematised way of working than a firefighter in a quiet rural area where they have a lower frequency of incident calls. This level of variability was less prominent in the sample group north European firefighters (including UK firefighters), which could indicate a more robust structure to the fire service's organisation and fire service operations.

This group also provided a smaller number of responses with negative weighting, with UK firefighters providing the greatest consistency in their responses. Conversely, the sample group of 'south European firefighters' demonstrated higher levels of variability in their responses and, comparatively, lower levels of responses with positive weighting. This could indicate a lower standard of overall organisational structure of these fire services when comparted to those in northern Europe (including the UK). However, it is interesting that some questions received similar responses across all sample groups. This was particularly pertinent to the questions that would demonstrate reliance in incident commanders and their decisions. It appears that across the board, reliance on incident commanders is not overtly high.

This could mean that:

- Their training is not adequate.
- Their decision making is felt not to be sound by operational staff.
- They do not possess character traits that provides reassurance to operational crews.
- They lack experience and credibility.
- They lack leadership skills.

Research conducted by Rapp and Wilson (2022) into factors that contribute to trustworthiness across levels of authority in wildland fire incident management teams found that there are 14 attributes that influence trust in incident commanders. These are:

1. Ability.
2. Initial impressions of competence and baseline qualifications.
3. Operationally sound decision-making.
4. Leadership.
5. Communication.
6. Benevolence.
7. Integrity.
8. Honest communication.
9. Personal humility.
10. Owning decisions.
11. Predictability.
12. Familiarity.
13. Reliability and availability.
14. Gender.

These aspects are refined into the below summary table and demonstrate how trusted and trustee interact with each other in accordance with the above 14 attributes. However, leadership studies have demonstrated that the skills required to be a good leader involve a combination of complex factors such as the innate ability to lead through specific character traits, a willingness to lead, to have the aptitude to learn leadership skills and to possess other attributes that instil trust in others such as knowledge and experience. Metz (2015: p 81) argues that leadership is four dimensional and must be developed through:

1. Explicit knowledge - the managerial aspect of leadership that can be learned from books.
2. Tacit knowledge - those aspects that cannot be learned from books and must be internalized by experience. Tacit knowledge can be developed through mentorship, feedback and reflection.
3. Emotional intelligence - empathy, integrity and self-awareness.
4. Ethical leadership - what embeds the development of morality and caretaking.

As leadership is a complex skill that requires a combination of interlinked qualities, it is reasonably assumed that not everyone in leadership roles within the fire service is equally good at being a leader or equally competent. As the integral function of the fire service requires workers (firefighters) and leaders (officers), it is likely that no matter how rigid training provisions for leadership roles may be, not every individual will possess the integral (and perhaps innate) characteristics to be a good leader. As the demand for people in this role is vital to the function of the fire service, then it is likely that emphasis is placed on the quantifiable variables to learning and teaching leadership such as through explicit knowledge rather than cultivating innate skills.

In this regard, it is likely that not all incident commanders will meet the holistic set of criteria for a 'good' leader and this could viably explain the commonalities in responses between fire services for this question. There are other cultural issues that might also be at play. Further qualitative research could be conducted to determine the validity of this assumption. It was suggested in Rapp et al (2022) that gender played an important role in the extent to which firefighters trusted incident commanders. This is a long-standing cultural issue in fire services that are deemed to be male dominated and institutionally sexist (HMICFRS, 2023). Cox (2012) and Brunsden et al. (2014) further explore how hegemonic masculinity in fire service culture can stifle emotional openness, reduce inclusivity, and diminish trust in leaders who do not conform to these norms. Other broader cultural values pertinent to emergency service occupations and in particular to the fire service also hold strong acceptance criteria. Such values would include elements such as experience, technical ability, fearlessness and aggression where people seen as weaker in these elements will generally not be accepted or valued as a firefighter or officer.

Research by Read and Parker (2008: p 133) concluded that their findings suggest that fire service personnel construct specific identities based on a range of values and qualities. The importance of the physical nature of their job is central but there is also a significant emphasis placed upon technical proficiency and the production of emotional control and restraint. It is primarily through the construction of physical, technical and emotional competence that self-worth is asserted. Read and Parker (2008) also suggest that firefighter identity is constructed around physicality, technical proficiency, and emotional restraint, often leaving little room for softer leadership traits.

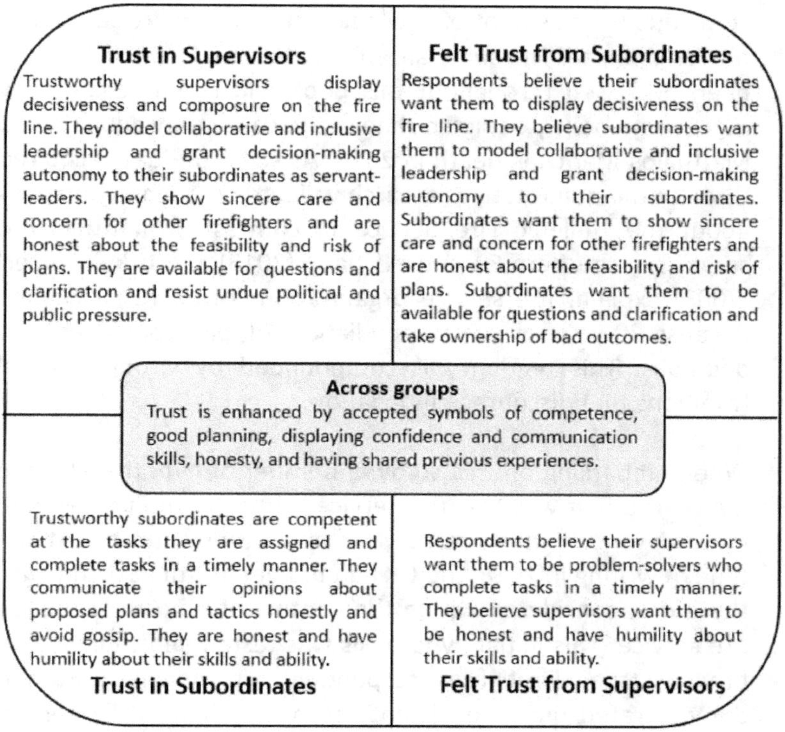

Figure 9 - Summary of important characteristics across trust referents

(Rapp and Wilson, 2022)

Greek firefighters provided, by far, the highest number of responses with negative weighting. In many cases, these responses were disproportionately higher than any other fire service in the category south-European firefighters. This would indicate that the organisation of the fire service in Greece across multiple thematic areas is less robust. This hypothesis was partially validated through this research although responses from other fire services in the sample group 'south European firefighters' were also problematic across numerous thematic areas. The differences between the Hellenic Fire Service and UK fire services were stark. This was also anticipated as part of this study due to the extent of available data on how fire services in the UK operate. The shortfalls of the Hellenic Fire service have been discussed throughout this book and nowhere are they more prominently depicted than in the case-study analysis into Matthaios Mantzios death in 2014. As such, the data acquired through the quantitative research validated certain hypotheses about the Hellenic Fire Service in contrast with other fire services in southern Europe. As Napp (2015) and Jick & Peiperl (2003) explain, fire service organisations often resist change because it is seen as a repudiation of past practices and decisions. This tendency is compounded by strong internal traditions that discourage innovation.

When embarking on this study, a number of hypotheses were made. The first was that fire services in northern Europe were more advanced than fire services in southern Europe. At the time of writing in 2018, the UK had not yet left the EU and was, therefore, considered within the context of a north European fire service. This initial hypothesis was tested and found to be broadly true. Firefighter respondents in northern Europe demonstrated more consistent responses than firefighters in the sample group 'south European firefighters'. This was particularly the case with UK firefighter respondents that provided the most consistent responses throughout the questionnaire.

The second hypothesis was that countries that suffered more as a result of the financial crisis i.e. the PIGS, were less likely to have a structured approach at an organisational and operational level due to a lack of resources and funds. This was undetermined, partially due to the relatively low numbers of responses but also because of other factors that could be at play such as corruption and other cultural influences that could equally serve to hinder fire service development. Although countries in the sample group 'south European firefighters' presented greater variability in their responses, of which a notably higher frequency of responses with negative weighting, it could not be determined without further research whether these shortcomings were as a result of limited access to required funds and resources or their mismanagement, the complexities arising from bureaucratic and administrative processes, poor culture or any combination of the above. The third hypothesis was that other fire services in southern Europe, despite experiencing the impacts of the financial crisis, were adapting better to change, and developing faster than the Hellenic Fire Service. This also appears to be true although the other fire services in the sample group 'south European firefighters' were not significantly better than the Hellenic Fire Service except in some areas. Responses acquired from Hellenic Fire Services firefighters provided the highest and most consistent frequency of responses with negative weighting.

As it was not possible to analyse causal factors behind responses for every fire service, the data for the Hellenic Fire Service and the UK fire services was selected and analysed in greater detail. These two sample groups provided the highest frequency of counter-opposite responses. UK firefighters provided the highest number of definitive responses and, therefore, it could be surmised that fire services in the UK have the most robust organisational and operational structure of all sample groups that took part in this study.

The high numbers of definitive responses means that these can be qualified through policies and procedures, and in this case, National Operational Guidance. This is why the low overall number of responses from UK firefighters is not of critical importance as these are broadly reflective of the status quo in UK fire services. There is also significant evidence that is publicly available to make sound assumptions about many of the responses provided in the questionnaire. Conversely, Greek firefighters provided the highest number of responses with a negative weighting meaning that the Hellenic Fire Service can be viewed comparatively as having the least robust organisational and operational structure of all sample groups. The reasons behind this were found to vary although the Matthaios Mantzios case study provides a detailed insight into the way the Hellenic Fire Service operates, even today.

This discussion chapter has analysed the key findings from the comparative study of European fire and rescue services, with a specific focus on the disparities between UK, north European, and south European fire services. The responses, when categorised into definitive, fluid, and perceptive types, revealed significant organisational and cultural divergences, particularly highlighting the structural rigidity and policy-driven approaches adopted by UK fire services as compared to the fragmented, often ambiguous practices observed in southern Europe—most notably within the Hellenic Fire Service.

A recurring theme in the analysis was the role of organisational learning—or lack thereof—in shaping both operational reliability and workforce confidence. Choo (1996) and Senge (1990) argue that many organisations suffer from learning disabilities, impeding their ability to unlearn obsolete practices and adopt better alternatives. This appears particularly relevant in the Hellenic Fire Service, where despite limited evidence of formalised standard operating procedures, many respondents perceived their existence—highlighting what Lupyan (2016) would describe as perception influenced by belief rather than evidence.

Leadership emerged as another critical variable influencing trust, operational consistency, and perceived safety. While UK fire services benefit from structured leadership frameworks supported by the National Operational Guidance and National Occupational Standards (CFOA, 2013), south European services were characterised by inconsistent leadership quality and limited faith in incident commanders. This aligns with the findings of Rapp and Wilson (2022), who emphasise that trust in leadership is shaped not only by operational ability, but also by attributes such as humility, predictability, and integrity—traits often underdeveloped in hierarchical or tradition-bound organisations.

Moreover, the cultural resistance to change, as described by Napp (2015) and Grimwood (2017), remains a powerful impediment in many fire services. When operational success is measured by tradition rather than evidence, organisations risk stagnation. As Hartin (2013) warns, "experience without understanding" undermines professional growth and may lead to unsafe practices that go unchallenged due to cultural inertia.

The chapter has also demonstrated that fire services with institutionalised mechanisms for learning from incidents, are better positioned to evolve, improve safety, and maintain public trust. Conversely, the absence of these mechanisms in countries like Greece limits systemic reform, perpetuates freelancing, and contributes to poor organisational memory—issues made tragically evident in the Matthaios Mantzios case study and sadly continue to this day.

In sum, the findings reinforce the hypothesis that fire services in northern Europe—particularly the UK—possess more structured and mature organisational systems. While financial constraints in southern Europe may explain some deficiencies, cultural, political, and structural factors appear equally significant. These insights form the foundation for the final chapter, which will consolidate the research findings, assess the validity of the study's initial hypotheses, and offer practical recommendations toward establishing a global baseline standard for fire service performance and quality assurance.

Chapter 7 - Conclusions

This study has explored the organisational structure of fire services and their capacity to respond to emergencies effectively, efficiently and safely. The principal focus has been between fire services in Northern Europe, the UK and fire services in Southern Europe. The study aimed to address a series of perceptions about the differences in how fire services operate between the two geographical regions. In general, it has been determined that the business of fire services is continuously evolving, and adapting in response to advances in technology, climate change, socio-economic factors, urbanisation etc. and the complexities that come as a result, is essential if fire services want to be effective in delivering their core business (Sample, 2022).

Although the principal functions of fire services for protecting life and property might remain the same, the way in which these are achieved have changed dramatically over the last couple of decades, not only through the acquisition of new and improved equipment but also in the methods in which operational tasks are addressed, as well as with the proliferation of stricter fire safety regulations in many countries (Burge, 2024). It has been demonstrated that experience alone will no longer suffice as a measure of effectiveness and that science also has a significant role to play in developing effective methods to emergency management (Svensson, 2018). Such developments need to be reflected through policies, continuous and consistent training, and operational procedures.

In the past, firefighting was as simple as putting lots of water on the fire and various aspects of firefighter prowess were learned intuitively. Firefighters would stay low in fire-involved structures because heat forced them to the ground (International Fire Service Training Association, 2018). They often didn't wear breathing apparatus because they felt it wasn't necessary or saw it as a hindrance (Evans, 2015). Events such as flashover did not become time-critical until the 1980s where a significant decrease in the time to flashover presented new risks to first arriving crews at structure fire incidents (Kerber, 2011). Donning personal protective equipment was optional in the past although its quality and performance was sub-standard to the equivalent PPE of the modern firefighter (Hasenmeier, 2008). Although this is broadly recognised, it was identified that not all fire services have the capacity, knowledge or will to adapt to new challenges.

The key themes that were identified and analysed in this study were in relation to the importance of:

- Crew consistency and staffing levels.
- Standard operating procedures and guidelines.
- Firefighter competency and training.
- A defined incident command structure.
- Personal protective equipment and radio communications.
- Firefighter health and safety.
- Lessons learned as a tool for improvement.

The Mantzios case-study demonstrates the likely outcome that can result from the failure to seriously address these matters at an organisational level. It is argued that the outcomes of such failures, especially when life and property is unnecessarily lost, are resultant of critical systemic failures that are often overlooked and where focus in placed on the intricacies of an event itself rather than the root causes.

Throughout this study, several conclusions were derived:

1. There is great diversity between fire services, even within the same country. The way end-goals are

achieved can vary significantly between fire services, without one method necessarily being better than another. However, even in cases where fire services appeared to have a more robust organisational structure, some methods were either more heavily focused on an experience-based approach without the supporting science to qualify those methods of work, or there was only a partial understanding of the supporting science, meaning that methods of work were incomplete.

2. There was a lack consensus between fire services. This means that not all matters pertinent to the safe and efficient work of fire service organisations are held in the same regard across the board. Fire services in northern Europe demonstrated commonalities in what they feel is important to them – in their core values. An example would be in reference to the appropriateness of PPE or the level of their training where north European firefighters (including UK firefighters) were more likely to agree that these are adequate than their south European counterparts. This could indicate that the extent of emphasis placed on such aspects at an institutional level varies (quite notably in some cases) between fire services. This issue was recurring across multiple thematic areas of the study where south European firefighters presented lower confidence levels in their responses. The reason this occurs cannot simply be due to the absence of adequate funding, and it is more likely that there are a series of complex and interrelated factors that serve to undermine these fire services. These will ultimately have to do with deeply rooted cultural issues (corruption, political involvement and influence, clientelist relationships, a general lack of understanding and resistance to change, an absence of meritocratic processes, the devaluation of science

where emphasis is based on experience alone and the extent of value given to 'the way we do things around here'). In this regard, an analysis of the situation in Greece has been used to demonstrate how these factors can affect the performance of a fire service, hinder its progress and development, and significantly increase the overall risk of fire and rescue operations.

3. Commitment to fundamental principles around fire service delivery can be overlooked where financial interests are at play, despite robust evidence demonstrating the importance of doing things in certain ways, like for example, the importance of crew sizes on fire engines. Although big data analysis is critical to the effective organisation of fire services, it can often be used to justify decisions made around pay and staff cuts where other qualitative information has been omitted or not considered. There are numerous examples from the UK and other continents where fire engines respond to incidents understaffed and fire stations have been closed in the name of cost-saving. In some cases, the evidence might be robust enough to justify some cost-saving initiatives but there is evidence to demonstrate that, in other cases, the outcomes can be disastrous if the data is analysed to satisfy specific objectives.

4. The way in which firefighters perceive problems is highly contextual. Very different problems may be viewed with the same weight depending on where in the world one is. In the UK, for example, a firefighter as part of a 5 person crew working beyond 5 hours at an emergency incident might feel the same level of discontent as firefighters in Greece working at a fire with a 2 person crew for twelve hours. Where firefighters in Northern Europe might be displeased that they have to wear soiled PPE on the fire engine

returning from an incident until they get back to the fire station where they will be provided with showers and new kit, firefighters in southern Europe might only have one set of PPE which they will need to retain for one or more shifts that is soiled with harmful particles. In these cases, soiled PPE is often cleaned at home in domestic family washing machines, thus increasing the risk of contamination to other family members from harmful chemicals such as PFAS. These problems, although different, are felt with the same level of dissatisfaction because the context and starting point is different.

5. Despite many fire services in northern Europe leading the way in the fire service world, there still appears to be an overreliance on experience-based methods of work or validation through repetitive action. This method of doing work often ignores scientific and other evidence demonstrating that existing methods are neither most effective nor most safe. There are commonalities here between north and south European fire services although these occurrences are perhaps less prominent in northern Europe.

6. It is difficult to measure the effectiveness of a fire service. Just because a fire has been extinguished this does not mean that this had been done well. Many positive incident outcomes have presented high-risk taking behaviour or employed outdated methods that are no longer efficient or safe. The only viable way to measure effectiveness is through the implementation of auditable processes. The United Kingdom has generated an independent auditing structure through His Majesty's Inspectorate of Constabularies and Fire and Rescue Services (HMICFRS). Without defined processes, it is almost impossible to measure effectiveness.

For fire services to be more effective, it is imperative that commitment to certain core values is practiced robustly (Correia, 2021). Initially, fire services must have an in-depth understanding of what they want to do and how they need to do it (McCormack, 2022). Knowing what to do is not always a simple task and would need to extend well beyond the fundamental goals of protecting life and property. Knowing what to do can only be reliably determined via the collection and analysis of multiple source data that is critical to the business of the fire and rescue service (HMICFRS, 2023). Once fire services know what they want to do, they must determine the most effective methods of doing it. The 'what do I want to do' conundrum would need to be addressed in the organisation's policy. This will form their mission statement which identifies their core values. The 'How' questions are then answered in the organisation's overarching strategy (Dudeney, 2018). This would have a similar layout to the four-year Integrated Risk Management Plans drafted by UK fire and rescue services in the UK that are area specific and data driven. The methodology for achieving organisational goals also includes aspects pertinent to training (how long and what training), the development of procedures or guidelines, ascertaining minimum staffing levels on fire engines, implementing a defined incident command structure and so on. In general, if a baseline standard is to be achieved among fire services, regardless of where they are in the world, it can be surmised that the following aspects must be covered:

- Multiple source data should be used to explore areas of significant interest to the fire and rescue service. These might have to do with demographic data like, for example:
    - The highest risk groups for residential fires which would derive a series of prevention

initiatives like the installation of fire detection in their homes and the provision of fire safety advice.

  - Causes of firefighter injuries in order to determine the adequacy of training or to identify any patterns that have gone undetected, like injuries at a specific incident type.
  - Response times against response targets.

- Training must be directly associated with the 'what do I want to do' conundrum. New firefighters should be trained sufficiently in order to satisfy the principal duties expected of them at emergency incidents. In the UK, training is intensive and the training curriculum is covered over a period of approximately 11 – 15 weeks where trainees learn about the equipment and procedures (WMFRS, 2024). They undertake series of theoretical and practical modules on structural firefighting, road traffic collisions, breathing apparatus, ladders, hoses and immediate emergency care (advanced first aid) etc. Training must incorporate drills against specific incident types via the implementation of SOPs so that they are learned perfectly[50].

  This standardisation of processes assists predictability, creates safe systems of work, it's efficient and if something goes wrong it is easier to identify where failures have occurred. Fire services should be able to determine the time that is required to prepare firefighters for supervised frontline duties and these

[50] Training for new firefighters in Greece is now 4 years and it was determined through the interviews that the curriculum is far from adequate in preparing firefighters for front line duties.

timeframes should be realistic in adequately addressing the 'want do I want to do' questions. Training should also be standardised so that existing firefighters have a reasonable understanding of the knowledge-base and skill-set that new firefighters bring to the job upon graduating from the fire academy. Training must also incorporate important lessons drawn from years of firefighter experience, as long as this can be substantiated evidentially. Learning from past experiences still serves as a fundamental basis for firefighter training (Rowan-White, 2025).

However, this alone is no longer sufficient and due to the sheer extent of publicly available research, fire services on a global scale do not have an excuse to omit evidence-based information from their training. The 'what do I want to do' question in respect of training applies across all roles and ranks. If it can be determined what a fire service wants to do and how they want to do it, then this should govern the type and content of training for each independent role at every rank. This will ensure that competence is maintained to high standard or at least satisfy the criteria set by the organisation in how they determine competency.

- In order to answer the 'how' questions, fire services must develop a suite of SOPS or SOGs that will determine who does what, when, where and how. Fire services in northern Europe and the UK have a numbering system depending on where one is seated on a fire engine. Each individual's number is associated with a series of tasks per incident type. In North America, incident response is task-based depending on the order of arrival and the type of vehicle in which a firefighter is riding (Grimwood, 2008). This systematisation of work assists in enhancing both efficiency and effectiveness and limits the risk of freelancing which is an inherently dangerous practice (Brunacini, 2010). This process will make it easier for a fire and rescue service to determine the number of firefighters they require per fire engine per incident type (crew consistency), how many fire engines or other vehicles they require per fire station, how many staff they require per fire station and ultimately how many firefighters and vehicles the fire and rescue service requires as a minimum to achieve its operational goals.

- The adequacy of Personal Protective Equipment is perhaps one of the most important factors to be considered by fire and rescue services. Adequate PPE does not only protect firefighters in the undertaking of their operational duties from immediate risks at emergency incidents. It also protects them from the long-term effects as a result of exposure to harmful substances found on their PPE that can lead to serious and terminal illnesses (Stec, 2020). In this respect, apart from the adequacy of the PPE itself, fire and rescue services should develop policies to ensure that the health and safety of firefighters is protected as best as reasonably practicable and any standards for the provision, use, cleaning and decontamination of PPE should consider the findings from relevant research.

- Radio communications can be considered as equally important to a firefighter's PPE, especially in interior firefighting operations or at incidents when direct line of sight with firefighting crews is not readily achievable (USFA, 2016). Fire and rescue services must ensure that firefighters are provided with a method of communicating incident critical information to a command and control post from where decisive action can be taken. This is most pertinent to firefighter maydays where radio communications can serve as a lifeline to frontline crews. All fire services should have a policy that governs the provision and use of radio communications.

- All fire and rescue services must have a defined incident command structure. This should be in a systematised format that creates specific tasks and accountabilities. Much like operational SOPs at task level, an incident command system should identify who does what, when, where and how at command and control level, and these accountabilities should be set against the functional areas of an emergency incident at both tactical and strategic level. Such tasks would determine a chain of command, sectorisation requirements, a clear reporting structure, accountability, resource management and other logistics etc (NFCC, 2023).

- All fire services must develop and foster a lessons-learned policy so that they are able learn from their mistakes and reoccurrences are avoided through better training. It is essential that fire services are open to constructive criticism in the name of safety, efficiency and effectiveness, and the prospect of development (Murray, 2024). It appears reasonable that any such investigations into notable incidents should be undertaken by an independent party in order to ensure objectiveness.

Fire services can achieve the same goals via different methods of equal robustness. There is usually no single method for addressing an operational task within a fire service context. As experiences are learned differently it can be assumed that fire services also learn differently from one another. This diversity is not necessarily problematic but it is often difficult to ascertain an international baseline standard. The learning of a fire service will depend on several factors. Some of these factors have to do with wider social and cultural influences that extend into various organisational areas of a fire and rescue service. It is difficult to challenge or change views where these are representative of dominant culture. Although this was apparent across all fire services, it was most prominent in south European fire services. It has already been determined that fire services in countries with a greater level of political, financial and social stability have a better aptitude towards learning (Kane et al., 2005). This is perhaps why fire services in Northern Europe and developed western countries are objectively better than fire services in southern Europe and other less developed countries.

Corruption and micro-politics are not isolated to south European fire and rescue services alone, these are also present in fire services in more developed countries. However, it appears that where core values are present and adhered to, influences from corruption and micro-politics are lesser than in countries that do not have or adhere to core values. In this respect, fire services without commitment or adherence to core values are more likely to be influenced by corruption and micro-politics. In conclusion, it appears from this research that the robustness of fire service organisations (if effectiveness could be measured on a numbering scale) is directly reflective of the dominant culture that is representative of their respective countries. In this regard, it appears that countries that have higher levels of disorganisation in their administrative processes and higher levels of corruption are more likely to demonstrate lower levels of organisation, effectiveness and progress of their fire services (amongst other public sector organisations). This is particularly the case for the Hellenic Fire Service where arguments put forward through the Mantzios case study analysis have been quantified through Greek firefighter responses to the study questionnaire. Greater context has been given to the socio-economic and political history of Greece that can assist in understanding cultural nuances underpinning critical factors affecting its fire service's ability to consolidate a robust organisational structure with meritocratic processes. These issues are far more challenging to address within the scope of this study alone.

Chapter 8 - Challenges and further Research

Addressing these issues in not a simple task. The logical process would be to create a global guidance document for fire service organisations. This guidance document would be similar to an ISO standard that sets general best practice guidelines to be adhered to in order to satisfy quality assurance objectives through robust management processes (Tari et al., 2012). These guidelines do not set the methodology for achieving specific goals but would assist organisations in preparing their management processes accordingly in order to meet the required standards. The general topics that would need to be addressed amongst others are, as aforementioned:

- Defining mission objectives and how the organisation aims to achieve them through training.
- The development of SOPs and SOGs, both of which need to incorporate scientific foundations.
- The determination of adequate crew sizes and staffing levels at organisation level.
- The incorporation of a robust radio communications policy with firefighter safety at the forefront.
- A defined incident command structure with specific accountabilities and tasks.
- A health and safety standard that will be developed to consider the longevity of firefighter health and their protection from terminal illness.
- The creation of an independent lessons learned policy after notable incidents.

His Majesty's Inspectorate of Constabulary and Fire and Rescue Services assessment process for fire and rescue services in England already provides a good example and starting of how such an initiative could be prepared and delivered. As per the

ISO standards, audits would need to be undertaken by independent industry specialists in order to ensure objectivity and fire services would need allocate staff to manage the delivery of set targets. In general, the UK has a comprehensive system to assess the performance of fire and rescue services, the targets they set to meet the needs of the communities they serve and so on, through the periodic Integrated Risk Management Plans. In general, UK fire services have a notable level of transparency in their processes which would assist in developing a similar standard for fire service quality assurance on an international level.

However, there are significant problems that would hinder such an initiative from occurring. Firstly, an independent audit process would expose weaknesses in the fire service organisation under assessment. Fire services that are not open to such exposure and are resistant to outside criticism would unlikely buy-in to such an initiative. Conversely, leaving such a process to be managed and audited internally would remove the important element of objectivity and potentially skew results at the expense of ensuring effectiveness (Fire Standards Board, 2024). Secondly, fire services need to be in a position to be able to adapt to changes required to address effectiveness and efficiency (Centre for Public safety and Excellence, 2020). This does not have to do with financial ability alone but also with the willingness to address these issues and the capability of the individuals tasked with these responsibilities. Although the financial constrains could be addressed through European funding programmes, if there is a generalised lack of understanding and competence within the organisation to address the importance of the aforementioned specific topics (training, crew size, the development of procedures etc.), it is unlikely that any actions towards achieving a higher standard of management around these aspects will be achieved.
Another constraint towards achieving these goals is the extent of authority held by decision makers and the extent of external

or internal influences. If decision makers within the fire service are limited in the extent to which they can affect change due to external political influences for example, then holistically addressing quality assurance within the organisation will be jeopardised (Launder & Perry, 2014). This could also be influenced by unions within the organisation that for whatever reason could obstruct or resist the implementation of new measures.

As such, although the solution for achieving a baseline standard across all fire service organisations via the implementation of good practice management initiatives appears simple, the complex socio-economic and cultural issues that are intricate to each country or fire service would present difficulties in their application. It appears that the resolution to addressing the issues put forward in this book are resultant of deeper cultural influences that are reflected through the organisational structure of fire and rescue services. In this respect, the resolution to these issues would be more pertinent to the social sciences and drivers around change in culture, rather than just teaching organisations how to better manage their core business. With this in mind, there are two prospects. The first is to teach the organisation what it needs to do in order to be most effective at achieving its goals for protecting life and property. It could then observe its gradual evolution over time until it has reached a point where it will be ready to adopt and maintain a systematic approach to providing service excellence using all the tools and information available to them, to support critical decision-making around mission objectives and approved methodologies for achieving them[51]. For this to occur, it would be vital for the right people to be in the right places to be able to influence and implement these changes.

---

[51] This appears to be the case with most fire services around the world where their development over time can be understood in a narrative of their history and progress, although it is evident that some organisations learn faster and better than others.

The second prospect would fall within the realms of risk acceptance. This means that, having knowledge of the complex and interrelated factors affecting the progress of a fire service and its resistance (or difficulty) to learn, such a circumstance leading to substandard performance would simply need to be accepted. The outcome of substandard performance has multiple long-standing effects on the life safety of the public as well as firefighters, on the successful salvage of property, on the morale and job satisfaction of firefighters and on mission outcomes in general. The problem is that the first two elements are not readily identifiable. This is because the public does not often understand the intricacies of fire and rescue operations in order to be in a position determine whether something could have been done better (Yamamoto, 2023).

Firefighters, on the other hand, without having the necessary knowledge and training, and without the use of defined procedures, will often have ignorance of potentially dangerous actions or updated methods to achieving specific goals more effectively (Pollack et al., 2017). Performance in such circumstances is most commonly assessed on outcome alone and this leads to a false sense of security by reaffirming specific actions that appeared successful at previous incidents and which are then recycled over and over at incidents to come.

Further research is required to determine a universal baseline standard for fire and rescue services i.e. what should be the minimum considerations required for effective organisation and fire and rescue operations. Such an initiative should set the fundamental principles that would be expected of any fire and rescue service that strives to provide service excellence. However, due to the diversity between nations and cultures, any recommendations should be set at a level that would be achievable for most and not necessarily require extensive funding. Good practice examples should also be given to serve as a process map that can be followed and adapted where required. These can be drawn from the existing work undertaken by the HMCFRS for UK fire and rescue services

Further research is also required to understand the correlation between dominant culture and the influence this has in a fire service context, with the scope of detecting critical areas that have a negative impact on the development of fire services and the standard of service delivery to the communities they serve. Although the findings of such research might not alone serve as a driver for change, especially if internal fire service culture is innately resistant to outside influences, the publication of findings would assist in developing awareness within the community as well as within the fire service and the government. Over time, such interventions might indeed serve as a driver for change, even if it is gradual. Promoting awareness empowers individuals or groups to formulate a better understanding of areas of interest, which in turn will adjust their expectations and sense of entitlement accordingly.

Bibliography and References

Acadis (2018). *Managing the Age-Old Volunteer vs. Career Firehouse Rivalry*. [online] Acadis. Available at: https://www.acadis.com/resource/managing-the-age-old-volunteer-vs-career-firehouse-rivalry/ [Accessed 10 Oct. 2021].

Adcroft, A. and Willis, R. (2005). The (un)intended Outcome of Public Sector Performance Measurement. *International Journal of Public Sector Management*, 18(5), pp.386–400. doi: https://doi.org/10.1108/09513550510608859.

Afonso, A., Schuknecht, L. and Tanzi, V. (2024). Public Sector efficiency: an International Comparison. *Handle.net*. [online] doi: http://hdl.handle.net/10419/152676.

Afzal, N. (2022). *Independent Culture Review of London Fire Brigade*. [online] *London Fire Brigade*. London: London Fire Brigade. Available at: https://www.london-fire.gov.uk/about-us/improving-our-workplace-culture/independent-culture-review/ [Accessed 23 Mar. 2024].

Agoston, R. (2018). The effects of global climate change on fire service Human resource view. *Procedia Engineering*, 211(211), pp.1–7. doi: https://doi.org/10.1016/j.proeng.2017.12.001.

Alexander, D. (1985). Culture and the Environment in Italy. *Environmental Management*, 9(2), pp.121–133. doi: https://doi.org/10.1007/bf01867112.

Alexander, D. (2002). From Civil Defence to Civil Protection and Back Again. *Disaster Prevention and Management: An International Journal*, 11(3), pp.209–213. doi: https://doi.org/10.1108/09653560210435803.

Alexander, D. (2005). Towards the Development of a Standard in Emergency Planning. *Disaster Prevention and Management: An International Journal*, 14(2), pp.158–175. doi: https://doi.org/10.1108/09653560510595164.

Alexander, D. (2006). The Globilization of Disaster: Trends, Problems and Dilemmas. *Journal of International Affairs Editorial Board (Spring and Summer 2006)*, [online] 59(2), pp.1–22. Available at: http://www.jstor.org/stable/24358424 [Accessed 20 Dec. 2017].

Alexander, D. (2010a). Civil Protection amid Disasters and Scandals. *Italian Politics*, [online] 26, pp.180–197. Available at: https://www.jstor.org/stable/43486785 [Accessed 2 Jul. 2018].

Alexander, D. (2010b). The Voluntary Sector in Emergency Response and Civil protection: Review and Recommendations. *International Journal of Emergency Management*, 7(2), p.151. doi: https://doi.org/10.1504/ijem.2010.033654.

Alexander, D. (2012). Models of Social Vulnerability to Disasters*. *RCCS Annual Review*, (4). doi: https://doi.org/10.4000/rccsar.412.

Alexander, D. (2015). *Disaster and Emergency Planning for Preparedness, Response, and Recovery. Oxford Research Encyclopedia of Natural Hazard Science*. Oxford University Press. doi: https://doi.org/10.1093/acrefore/9780199389407.013.12.

Alexander, D. (2017). Corruption and the Governance of Disaster Risk. *Oxford Research Encyclopedia of Natural Hazard Science*. [online] doi: https://doi.org/10.1093/acrefore/9780199389407.013.253.

Alexander, D. (2020). Emergency Management Systems after Disastrous Earthquakes Using Optimization methods: a Comprehensive Review. *Advances in Engineering Software*, 149(3), p.102885. doi: https://doi.org/10.1016/j.advengsoft.2020.102885.

Alexander, D. (2010c). The L'Aquila Earthquake of 6 April 2009 and Italian Government Policy on Disaster Response. *Journal of Natural Resources Policy Research*, 2(4), pp.325–342. doi: https://doi.org/10.1080/19390459.2010.511450.

Alexander, D. (2013). Resilience and Disaster Risk reduction: an Etymological Journey. *Natural Hazards and Earth System Sciences Discussions*, [online] 1(2), pp.1257–1284. doi: https://doi.org/10.5194/nhessd-1-1257-2013.

Alexander, D. (2014). Social Media in Disaster Risk Reduction and Crisis Management. *Science and Engineering Ethics*, 20(3), pp.717–733. doi: https://doi.org/10.1007/s11948-013-9502-z.

Ambinder, M. (2010). *The Fires This Time: Joe Flood on Managing New York City*. [online] The Atlantic. Available at: https://www.theatlantic.com/politics/archive/2010/05/the-fires-this-time-joe-flood-on-managing-new-york-city/56682/ [Accessed 24 Nov. 2018].

Andrews, R., Ashworth, R. and Meier, K.J. (2014). Representative Bureaucracy and Fire Service Performance. *International Public Management Journal*, 17(1), pp.1–24. doi: https://doi.org/10.1080/10967494.2014.874253.

Argote, Linda & Miron-Spektor, Ella. (2011). Organizational Learning: From Experience to Knowledge. Organization Science. 22. 10.2307/41303106.

Arvantinis T. (2013). *The Price of Rouspheti in the Employment Market*. [online] Capital.gr. Available at: https://www.capital.gr/arthra/1834092/to-kostos-ton-rousfetologikon-proslipseon/ [Accessed 11 Apr. 2021].

Associated Press (2008). *Greek Prosecutor Files Charges over Alleged Siemens Corruption Scandal - International Herald Tribune*. [online] web.archive.org. Available at: https://web.archive.org/web/20081012074516/http://www.iht.com/articles/ap/2008/07/01/europe/EU-Greece-Corruption-Siemens.php [Accessed 17 Feb. 2019].

Avsec, R. (2013a). *Changing the Fire Service Culture – Fire & EMS Leader Pro*. [online] www.fireemsleaderpro.org. Available at: https://www.fireemsleaderpro.org/2014/07/31/changing-fire-service-culture/ [Accessed 15 May 2018].

Avsec, R. (2013b). *What Is the Fire Service Culture? – Fire & EMS Leader Pro*. [online] Fireemsleaderpro.org. Available at: https://www.fireemsleaderpro.org/2013/12/06/what-is-the-fire-service-culture/ [Accessed 11 Apr. 2021].

Avsec, R. (2017). *Rapid Response vehicles: the Answer to Downsizing Fire Apparatus*. [online] FireRescue1. Available at: https://www.firerescue1.com/fire-products/fire-apparatus/articles/rapid-response-vehicles-the-answer-to-downsizing-fire-apparatus-Bo2XJbnsFRivswRt/ [Accessed 11 Apr. 2019].

Bakos, F. (2014). *Economic Crisis in Greece*. Dissertation. Technological Academic Institute of Piraeus, Department of Accounting and Finance.

Ballam, E. (2015). *Firehouse World: Fire Service Tradition Is Challenged by Progress*. [online] Firehouse.com. Available at: https://www.firehouse.com/home/news/12039411/firehouse-world-fire-service-tradition-is-challenged-by-progress-topic-at-moderated-session [Accessed 26 Apr. 2020].

Barowy, A.M., Notarianni, K., Averill, J.D., Moore-Merrell, L. and Santos, R. (2010). Community Risk Assessment Modeling: Results of Residential Fireground Field Experiments. *NIST*, [online] pp.1875–1886. Available at: https://www.nist.gov/publications/community-risk-assessment-modeling-results-residential-fireground-field-experiments?pub_id=905814 [Accessed 8 Aug. 2024].

Bartlett, W. and Uvalic, M. (2013). The Social Consequences of the Global Economic Crisis in Southeast Europe - LSE Research Online. *Lse.ac.uk*. [online] doi:http://eprints.lse.ac.uk/120489/1/Bartlett_social_consequences_of_the_global_economic_crises_published.pdf.

BBC News (2019). *Paris fire: Ten Dead and Many Injured at Apartment Block*. [online] Bbc.com. Available at: https://www.bbc.com/news/world-europe-47126440 [Accessed 27 Jan. 2020].

Beaton, R. and Ricks, D. eds., (2013). *The Making of Modern Greece*. USA: Ashgate Publishing Ltd.

Beeby, M. and Booth, C. (2000). Networks and inter-organizational learning: a critical review. *The Learning Organization*, 7(2), pp.75–88. doi: https://doi.org/10.1108/09696470010316260.

Belias, D. (2014). *Leadership and Job Satisfaction - a Review.* [online] *European Scientific Journal.* Available at: https://www.researchgate.net/publication/261570285 [Accessed 1 Oct. 2023].

Benny, L. (2019). *Review of Emergency Responder Standard Operating Procedures/Guidelines | NFPA.* [online] Nfpa.org. Available at: https://www.nfpa.org/education-and-research/research/fire-protection-research-foundation/projects-and-reports/review-of-emergency-responder-standard-operating-procedures-guidelines-sop-sog [Accessed 22 Apr. 2021].

Bianchi, Claudio & Giarelli, L & Grandi, G & Brollo, Alessandro & Ramani, L & Zuch, C. (1997). Latency periods in asbestos-related mesothelioma of the pleura. European journal of cancer prevention: the official journal of the European Cancer Prevention Organisation (ECP). 6. 162-6.

Brinchmann BC, Bugge MD, Nordby KC, Alfonso JH. Firefighting and melanoma, epidemiological and toxicological associations: a case report. Occup Med (Lond). 2022 Feb 22;72(2):142-144. doi: 10.1093/occmed/kqab183. PMID: 35064261; PMCID: PMC8863077.

Birkmann, J., Cardona, O.D., Carreño, M.L., Barbat, A.H., Pelling, M., Schneiderbauer, S., Kienberger, S., Keiler, M., Alexander, D., Zeil, P. and Welle, T. (2013). Framing vulnerability, risk and societal responses: the MOVE framework. *Natural Hazards*, 67(2), pp.193–211. doi: https://doi.org/10.1007/s11069-013-0558-5.

Birkmann, J., Kienberger, S. and Alexander, D.E. (2014). Introduction Vulnerability: a key determinant of risk and its importance for risk management and sustainability. *Elsevier eBooks*, pp.ix–xiii. doi: https://doi.org/10.1016/b978-0-12-410528-7.02001-4.

Bisschop, R., Willstrand, O., Amon, F. and Rosengren, M. (2019). *Fire Safety of Lithium-Ion Batteries in Road Vehicles. RISE Research Institutes of Sweden*, doi: https://doi.org/10.13140/RG.2.2.18738.15049.

Boussalham, H. (2018). The Consequences of Corruption on Economic Growth in Mediterranean Countries: Evidence from Panel Data Analysis. *www.preprints.org*, [online] 3. doi: https://doi.org/10.20944/preprints201802.0065.v3.

Brandweeracademie (2019). 11th International Fire Instructors Workshop (IFIW). A Report. Arnhem: Instituut Fysieke Veiligheid

Brandweeracademie (2019). *Recruitment and retention of volunteers in European Fire Services*. Arnhem: Instituut Fysieke Veiligheid.

Brauer, B.R. (2016). *The Fire and Emergency Services Culture: Can It Be Changed? - Fire Engineering: Firefighter Training and Fire Service News, Rescue*. [online] https://www.fireengineering.com/. Available at: https://www.fireengineering.com/firefighting/the-fire-and-emergency-services-culture-can-it-be-changed/#gref [Accessed 11 Apr. 2021].

Brewer, J.D. (2000). *Ethnography*. Buckingham; Philadelphia, PA: Open University Press.

Brislin, R.W., Lonner, W.J. and Thorndike, R.M. (1973). *Cross-cultural Research Methods*. New York: J. Wiley.

Brunacini, A. (1996). *StackPath*. [online] www.firehouse.com. Available at: https://www.firehouse.com/community-risk/article/10545449/always-be-nice-treat-everyone-with-respect-kindness-patience-and-consideration-alan-brunacini [Accessed 10 Oct. 2021].

Brunacini, A. (2010). *Freelancing Isn't Free - Fire Engineering: Firefighter Training and Fire Service News, Rescue*. [online] Fireengineering.com. Available at: https://www.fireengineering.com/firefighting/freelancing-isn-rsquo/#gref [Accessed 13 Jun. 2020].

Bryant, A. (2018). *How the Jaws of Life Became a Tool to Snatch You from the Jaws of Death*. [online] FireRescue1. Available at: https://www.firerescue1.com/fire-products/extrication-tools-cutters-and-spreaders/articles/how-the-jaws-of-life-became-a-tool-to-snatch-you-from-the-jaws-of-death-cZf0FflBNVLiTVuS/ [Accessed 11 Apr. 2019].

Buckman III, J. (2017). *13 Things That Make Fire Department Leaders Great*. [online] FireRescue1. Available at: https://www.firerescue1.com/exclusives-1/articles/13-things-that-make-fire-department-leaders-great-6yNcdRrWElI1y8Zl/ [Accessed 18 Jun. 2018].

Bucks Fire and Rescue (2020). *On-Call Availability – Responding to Incidents*. [online] Buckinghamshire Fire & Rescue Service. Available at: https://bucksfire.gov.uk/on-call-availability-responding-to-incidents/#:~:text=Our%20response%20model%20provides%20us [Accessed 13 Aug. 2021].

Bureau of Fire Prevention, C.D.U. (2021). *NYC Fire Code Guide Code Development Unit Bureau of Fire Prevention*. [online] Available at: https://www.nyc.gov/assets/fdny/downloads/pdf/business/nyc-fire-code-guide.pdf [Accessed 13 Jan. 2022].

Burge, S. (2024). *List of Firefighting Equipment*. [online] International Fire & Safety Journal. Available at: https://internationalfireandsafetyjournal.com/list-of-firefighting-equipment [Accessed 24 Apr. 2024].

Burke, E. (1997). Competence in command: Recent R&D in the London Fire Brigade. *Journal of Managerial Psychology*, 12(4), pp.261–279. doi: https://doi.org/10.1108/02683949710174865.

Butler, P.C., Bowers, A., Smith, A.P., Cohen-Hatton, S.R. and Honey, R.C. (2021). Decision Making Within and Outside Standard Operating Procedures: Paradoxical Use of Operational Discretion in Firefighters. *Human Factors: The Journal of the Human Factors and Ergonomics Society*, 65(7), p.001872082110418. doi: https://doi.org/10.1177/00187208211041860.

Cairns, G., Wright, G., Van der Heijden, K., Bradfield, R. and Burt, G. (2006). Enhancing Foresight between Multiple agencies: Issues in the Use of Scenario Thinking to Overcome Fragmentation. *Futures*, 38(8), pp.1010–1025. doi: https://doi.org/10.1016/j.futures.2005.12.020.

Cambanis, T. (2014). *Why Can't Greece Shake Its Corruption problem? - the Boston Globe*. [online] BostonGlobe.com. Available at: https://www.bostonglobe.com/ideas/2014/08/22/why-can-greece-shake-its-corruption-problem/6nwZglU9cmn70xg2x4sobM/story.html [Accessed 14 Nov. 2019].

Campbell, R. and Hall, S. (2023). *United States Firefighter Injuries*. [online] Nfpa.org. Available at: https://www.nfpa.org/education-and-research/research/nfpa-research/fire-statistical-reports/firefighter-injuries-in-the-united-states?l=0 [Accessed 10 Oct. 2023].

Carter, H. (2014). *Position Your Fire Department for Success*. [online] Firehouse.com. Available at: https://www.firehouse.com/leadership/professional-associations-services/article/11403803/how-to-create-a-successful-fire-department-in-the-21st-century [Accessed 11 Apr. 2021].

Carvalho, J., Fernandes, M., Lambert, V. and Lapsley, I. (2006). Measuring Fire Service performance: a Comparative Study. *International Journal of Public Sector Management*, 19(2), pp.165–179. doi: https://doi.org/10.1108/09513550610650428.

Centre for Public Safety Excellence (2020). *Fire and Rescue Service Leaders at the Core of Better Communities ® Fire and Emergency Services*. [online] Available at: https://www.cpse.org/wp-content/uploads/2020/07/21st-Century-Fire-and-Emergency-Services-White-Paper-Final-07.15.20.pdf [Accessed 4 Jun. 2024].

Chalaris, M. (2008). Health and Safety in the Fire Service: First Principles of Risk Assessment for Activities Presenting Inherent Risks. *Fire Service Inspector*, [online] (131 & 132). Available at: https://www.calameo.com/books/001962547f8d1d6ebaf77 [Accessed 22 Apr. 2018].

Cheshire Fire and Rescue Service (2019). *Cheshire Fire & Rescue Service Vehicles*. [online] Cheshire Fire & Rescue Service. Available at: https://www.cheshirefire.gov.uk/about-us/local-fire-stations/vehicles/ [Accessed 13 Aug. 2024].

Chesire Fire Brigades Union (2018). *Response to Cheshire Fire and Rescue Integrated Risk Management Plan 2018 - 19 and Interim Review of the Management of the Service*. Cheshire: Cheshire Fire Brigades Union.

Chief Fire Officers Association (2013). *Death in the Workplace Guidance for United Kingdom Fire and Rescue Services*. Staffordshire: CFOA Publications Ltd (CPL).

Chief Fire Officers Association (2013). *Command Training, Assessment, and Qualification Fire and Rescue Service Guidance*. Staffordshire: Chief Fire Officers Association.

Chief Fire Officers Association (2020). *Fighting Fires or Firefighting: The Impact of Austerity on English Fire and Rescue Services*. Staffordshire: Chief Fire Officers Association.

Chief Fire Officers Association (n.d.). *Investigative Guidance Firefighter Fatality in the Workplace*. [online] Available at: https://library.college.police.uk/docs/APPREF/firefighter-fatality-in-the-workplace.pdf [Accessed 9 Aug. 2024].

Chionis, C. (2015). *Sworn Administrative Examination into the Fatality of Sergeant Mathaios Mantzios*. Athens: Hellenic Fire Service.

Choo, C.W. (1996). The Knowing organization: How Organizations Use Information to Construct meaning, Create Knowledge and Make Decisions. *International Journal of Information Management*, [online] 16(5), pp.329–340. doi: https://doi.org/10.1016/0268-4012(96)00020-5.

Chung, S., Smith, M., Jin, A., Hogewood, L., Maksim Kitsak, Cegan, J. and Igor Linkov (2024). Access to Emergency Services: a New York City Case Study. *Transportation Research Interdisciplinary Perspectives*, 25, pp.101111–101111. doi: https://doi.org/10.1016/j.trip.2024.101111

Civil Protection Department (2018). *Legislative Decree No. 1 of January 2, 2018: Civil Protection Code*. [online] Protezionecivile.gov.it. Available at: https://www.protezionecivile.gov.it/en/normativa/legislative-decree-no--1-of-january-2--2018--civil-protection-code/ [Accessed 13 Aug. 2021].

Claessens, S., Ayhan, K.M., Laeven, L. and Valencia, F. (2024). Understanding Financial Crises: Causes, Consequences, and Policy Responses. *Handle.net*. [online] doi:http://hdl.handle.net/10419/108616.

Clean Water Action (2020). *PFAS-Containing Firefighting Foams*. [online] Clean Water Action. Available at: https://cleanwater.org/pfas-containing-firefighting-foams#:~:text=AFFF%20are%20synthetic%20PFAS%2Dcontaining [Accessed 26 Mar. 2024].

Clogg, R. (2016). *A Concise History of Greece*. Cambridge: Cambridge University Press.

CNN Greece (2019). *Σε Βραχυκύκλωμα Οφείλεται Η Φωτιά Στο Πλοίο 'Ελευθέριος Βενιζέλος'*. [online] CNN.gr. Available at: https://www.cnn.gr/ellada/story/189682/se-vraxykykloma-ofeiletai-i-fotia-sto-ploio-eleytherios-venizelos [Accessed 13 Aug. 2020].

Collins, A.E., Samantha, J., Manyena, B. and Jayawickrama, J. (2014). *Hazards, Risks, and Disasters in Society*. [online] *Google Books*. Elsevier. Available at: https://books.google.gr/books?hl=en&lr=&id=GlxzAwAAQBAJ&oi=fnd&pg=PP1&dq=Hazards+and+Disasters+Series+Hazards [Accessed 2 Aug. 2019].

Correia, T. (2021). *Character, Attitudes, and Values: Defining the 'Good' Firefighter*. [online] Fire Engineering: Firefighter Training and Fire Service News, Rescue. Available at: https://www.fireengineering.com/firefighter-training/character-attitudes-and-values-defining-the-good-firefighter [Accessed 24 Apr. 2024].

Costas Azariadis and Ioannides, Y.M. (2017). Thinking about Corruption in Greece. *The MIT Press eBooks*. doi: https://doi.org/10.7551/mitpress/9780262035835.003.0013.

Cox, A.T. (2012). *The Significance of Fire Service Culture an Impediment to Effective Leadership in the Homeland Security Environment*. Postgraduate Thesis.

Daniels, R.D., Kubale, T.L., Yiin, J.H., Dahm, M.M., Hales, T.R., Baris, D., Zahm, S.H., Beaumont, J.J., Waters, K.M. and Pinkerton, L.E. (2013). Mortality and cancer incidence in a pooled cohort of US firefighters from San Francisco, Chicago and Philadelphia (1950–2009). *Occupational and Environmental Medicine*, [online] 71(6), pp.388–397. doi: https://doi.org/10.1136/oemed-2013-101662.

Danopoulos, C.P. (2015). Accountability and the Quality of Democracy in Greece. *Mediterranean Quarterly*, 26(4), pp.110–131. doi: https://doi.org/10.1215/10474552-3425211.

Dearden, L. (2022). *Patients Taken to Hospital in Fire Engines as Firefighters Respond to Rocketing Medical Calls*. [online] https://www.independent.co.uk/. Available at: https://www.independent.co.uk/news/uk/home-news/ambulance-crisis-firefighters-nhs-b2147909.html [Accessed 13 Sep. 2022].

Deiner, C. (2017). Vehicle Extrication with Limited Staffing Levels: Surviving in a cut-back environment. *Fire and Rescue International,* [online] 4(6). Available at: https://www.frimedia.org/uploads/1/2/2/7/122743954/colin-vol4-no6.pdf [Accessed 24 Mar. 2024].

Delano Smith, C. and Bowman, J.S. (2024). Greece - Ottoman Rule, Resistance, Revolution | Britannica. In: *Encyclopedia Britannica*. [online] Available at: https://www.britannica.com/place/Greece/Resistance-to-Ottoman-rule#toc26407 [Accessed 8 Aug. 2024].

Della Porta, D. and Keating, M. (2008). *Approaches and Methodologies in the Social Sciences: a Pluralist Perspective*. Leiden: Cambridge University Press.

Dennison, K.J., Mullineaux, D.R., Yates, J.W. and Abel, M.G. (2012). The Effect of Fatigue and Training Status on Firefighter Performance. *Journal of Strength and Conditioning Research*, 26(4), pp.1101–1109. doi: https://doi.org/10.1519/jsc.0b013e31822dd027.

Department for Communities and Local Government (2012a). *Fire and Rescue National Framework for England*. London: Department for Communities and Local Government.

Department for Communities and Local Government (2012b). *Fire and Rescue Service Operational Guidance Operational Risk Information*. London: Department for Communities and Local Government.

Department for Communities and Local Government (2013). *Fire and Rescue Authorities Health, Safety and Welfare Framework for the Operational Environment*. London: Department for Communities and Local Government.

Department for Communities and Local Government (2014). *Addendum to the Fire and Rescue National Framework for England. Firefighter Fitness*. London: Department for Communities and Local Government.

Deutscher Feuerwehr Vebrand (2021). *Statistik*. [online] Feuerwehrverband. Available at: https://www.feuerwehrverband.de/presse/statistik/ [Accessed 7 Jul. 2022].

Devi, R. and Shaik, N. (2020). *Evaluating Training & Development Effectiveness -A Measurement Model*. [online] Available at: https://www.researchgate.net/publication/340875030_Evaluating_training_development_effectiveness_-A_measurement_model?enrichId=rgreq-0e3fa598ae81908033ec376a66a2177a-XXX&enrichSource=Y292ZXJQYWdlOzM0MDg3NTAzMDtBUzo4ODM1NTQyMDAzNDI1MjlAMTU4NzY2NzE0NTk5MA%3D%3D&el=1_x_2&_esc=publicationCoverPdf [Accessed 30 Aug. 2020].

Doyle, B. (2011). *About Lincolnshire*. [online] Lincolnshire.org. Available at: https://lincolnshire.org/about-lincolnshire/ [Accessed 6 Jun. 2019].

Efthimiou, N., Psomiadis, E. and Panagos, P. (2019). Fire Severity and Soil Erosion Susceptibility Mapping Using multi-temporal Earth Observation data: the Case of Mati Fatal Wildfire in Eastern Attica, Greece. *CATENA*, 187, p.104320. doi: https://doi.org/10.1016/j.catena.2019.104320.

Elkjaer, B. (2004). Organizational Learning. *Management Learning*, 35(4), pp.419–434. doi: https://doi.org/10.1177/1350507604048271.

Emerald Article (2010). Enhanced organizational learning. *Development and Learning in Organizations: An International Journal*, 24(1), pp.27–29. doi: https://doi.org/10.1108/14777281011010505.

Emergency Management Services International, Inc (n.d.). *History of ICS – EMSI.* [online] Emergency Management Services International, Inc. Available at: https://www.emsics.com/history-of-ics/ [Accessed 15 May 2019].

Emerson, R.M. (1964). Power-Dependence Relations: Two Experiments. *Sociometry*, 27(3), p.282. doi: https://doi.org/10.2307/2785619.

Eslamzadeh, M.K., Grilo, A. and Espadinha-Cruz, P. (2022). A Framework for Resource Allocation in Fire Departments: a Structured Literature Review. *Fire*, 5(4), p.109. doi: https://doi.org/10.3390/fire5040109.

Essex County Fire and Rescue Service (2015). *Independent Cultural Review of Essex County Fire and Rescue Service.* Essex: Essex County Fire and Rescue Service.

Essex County Fire & Rescue Service (2016). *Integrated Risk Management Plan 2016–2020*. Essex: Essex County Fire and Rescue Service.

European Parliament (2014). *Impact of the Economic Crisis on Social, Economic and Territorial Cohesion of the European Union.* Brussels: EU Policy Department.

Eurostat (2023). *EU Had Almost 360 000 Professional Firefighters in 2022 - Products Eurostat News - Eurostat.* [online] ec.europa.eu. Available at: https://ec.europa.eu/eurostat/web/products-eurostat-news/w/ddn-20230807-1 [Accessed 22 Oct. 2023].

Evans, R. (2015) 'The Evolution of SCBA and Training Innovations', *Fire Engineering.* Available at: https://www.fireengineering.com/firefighter-training/evolution-scba-training-innovations/ (Accessed: 24 April 2025).

Fahy, R.F., LeBlanc, P.R. and Molis, J.L. (2018). *NFPA: Firefighter Fatalities in the U.S. 2017 - Firefighter Nation: Fire Rescue - Firefighting News and Community.* [online] Firefighternation.com. Available at: https://www.firefighternation.com/leadership/nfpa-firefighter-fatalities-in-the-u-s-2017/#gref [Accessed 25 Jul. 2021].

Falmouth Packet (2015). *Fighting Fire on St Michael's Mount to Be Tested.* [online] Falmouth Packet. Available at: https://www.falmouthpacket.co.uk/news/11868316.fighting-fire-on-st-michaels-mount-to-be-tested/ [Accessed 22 Apr. 2020].

Federal Emergency Management Agency (2021). *Fire Data Analysis Handbook.* [online] Available at: https://www.usfa.fema.gov/downloads/pdf/publications/fa-266.pdf.

Feliu, F. and Josep Maria Nadal (2016). *Constructing Languages.* Amsterdam: John Benjamins Publishing Company.

Fennell, D. and Britain., G. (1988). *Investigation into the King's Cross Underground Fire.* London: H.M.S.O.

Fernandez, A.R., Treichel, A., Myers, J.B., Bourn, S.S., Crowe, R.P. and Gardner, B. (2023). Evaluating Firefighter On-Scene Decontamination Practices Using a National Fire Records Management System. *Journal of Occupational and Environmental Medicine*, [online] 65(11), p.931. doi: https://doi.org/10.1097/JOM.0000000000002927.

Fiol, C.M. and Lyles, M.A. (1985). Organizational Learning. *The Academy of Management Review*, [online] 10(4), p.803. doi: https://doi.org/10.2307/258048.

Fire Brigades Union (1996). *Fatal Accident Investigation. 14 Zephaniah Way, Blaina Gwent, 1st February 1996: Report and Conclusions*. Gwent: Fire Brigades Union.

Fire Brigades Union (2004). *Findings of the FBU Health and Safety Investigation into the Fatal Fire Which Occurred at Bethnal Green Road on 20 July 2004*. London: Fire Brigades Union.

Fire Brigades Union (2008). *In the Line of duty. Firefighter Deaths in the UK since 1978. a Report by the Labour Research Department (LRD) for the Fire Brigades Union.* Surrey: Fire Brigades Union.

Fire Brigades Union (2009). *Fatal Accident Investigation Report into the Death of Firefighter Ewan Williamson at the Balmoral Bar Dalry Road, Edinburgh*. Surrey: Fire Brigades Union.

Fire Brigades Union (2010). *It's about Time Why Emergency Response Times Matter to Firefighters and the Public*. Surrey: Fire Brigades Union.

Fire Brigades Union (2017). *Fire Service to Increase Crew Sizes after Union Outcry*. [online] Fire Brigades Union. Available at: https://www.fbu.org.uk/news/2017/09/22/fire-service-increase-crew-sizes-after-union-outcry [Accessed 6 Jun. 2019].

Fire Brigades Union (2018). *The Grenfell Tower Fire: Background to an Atrocity. The Fire Safety Regime and the Fire and Rescue Service.* Surrey: Fire Brigades Union.

Fire Engineering Staff (1995). *Tragedy in a Residential High-Rise, Memphis, Tennessee - Fire Engineering: Firefighter Training and Fire Service News, Rescue*. [online] https://www.fireengineering.com/. Available at: https://www.fireengineering.com/firefighting/tragedy-in-a-residential-high-rise-memphis-tennessee/#gref [Accessed 10 Oct. 2021].

Fire Engineering Staff (2002). *How Firefighters Learn on the Job - Fire Engineering: Firefighter Training and Fire Service News, Rescue*. [online] Fireengineering.com. Available at: https://www.fireengineering.com/firefighting/how-firefighters-learn-on-the-job/#gref [Accessed 10 Aug. 2022].

Fire Engineering Staff (2005). *ORGANIZATIONAL CULTURE IN THE FIRE SERVICE - Fire Engineering: Firefighter Training and Fire Service News, Rescue*. [online] Fireengineering.com. Available at: https://www.fireengineering.com/leadership/organizational-culture-in-the-fire-service/#gref [Accessed 11 Apr. 2021].

Fire Engineering Staff (2008). *Engine Company Assignments - Fire Engineering: Firefighter Training and Fire Service News, Rescue.* [online] Fireengineering.com. Available at: https://www.fireengineering.com/leadership/engine-company-assignments/#gref [Accessed 10 Aug. 2022].

Fire Engineering Team (2008). *Fire Service Court: Implementing SOGs - Fire Engineering: Firefighter Training and Fire Service News, Rescue*. [online] Fireengineering.com. Available at: https://www.fireengineering.com/leadership/legal/fire-service-court-implementing-sogs/#gref [Accessed 10 Aug. 2022].

Fire Engineering Team (2009). *Fire Service Court: SOGs versus SOPs - Fire Engineering: Firefighter Training and Fire Service News, Rescue*. [online] https://www.fireengineering.com/. Available at: https://www.fireengineering.com/leadership/legal/fire-service-court-sogs-versus-sops/#gref [Accessed 10 Aug. 2022].

Fire Engineering Team (2016). *Traditions and Progress in the Fire Department - Fire Engineering: Firefighter Training and Fire Service News, Rescue*. [online] Fireengineering.com. Available at: https://www.fireengineering.com/leadership/traditions-and-progress-in-the-fire-department/#gref [Accessed 10 Aug. 2022].

Fire Rescue 1 (2017). *The 10 Traits All Great Firefighters Have.* [online] FireRescue1. Available at: https://www.firerescue1.com/career-1/articles/the-10-traits-all-great-firefighters-have-GqyKt2N3QXqFs5wl/ [Accessed 18 Mar. 2018].

Firefighter Nation (2019). *Can We Do More with Less? - Firefighter Nation: Fire Rescue - Firefighting News and Community*. [online] www.firefighternation.com. Available at: https://www.firefighternation.com/firerescue/can-we-do-more-with-less/#gref [Accessed 10 Oct. 2021].

Firefighter Nation Staff (2009). *Human Factors & the Fire Service - Firefighter Nation: Fire Rescue - Firefighting News and Community.* [online] Firefighternation.com. Available at: https://www.firefighternation.com/firerescue/human-factors-the-fire-service/#gref [Accessed 11 Jul. 2022].

Firefighter Nation Staff (2010). *Managing Your Crew Effectively on the Fireground - Firefighter Nation: Fire Rescue - Firefighting News and Community.* [online] Firefighternation.com. Available at: https://www.firefighternation.com/firerescue/managing-your-crew-effectively-on-the-fireground/#gref [Accessed 11 Aug. 2024].

Firefighter Nation Staff (2013a). *Communication & Personnel Accountability for ICs - Firefighter Nation: Fire Rescue - Firefighting News and Community.* [online] Firefighternation.com. Available at: https://www.firefighternation.com/firerescue/communication-personnel-accountability-for-ics/#gref [Accessed 10 Aug. 2022].

Firefighter Nation Staff (2013b). *Influencing Cultural Change - Firefighter Nation: Fire Rescue - Firefighting News and Community.* [online] Firefighternation.com. Available at: https://www.firefighternation.com/firerescue/influencing-cultural-change/#gref [Accessed 11 Apr. 2021].

Firefighter Nation Staff (2014). *Big Data in the Fire Service: a Primer - Firefighter Nation: Fire Rescue - Firefighting News and Community.* [online] Firefighternation.com. Available at: https://www.firefighternation.com/firerescue/big-data-in-the-fire-service-a-primer/#gref [Accessed 11 Aug. 2021].

Firefighter Nation Staff (2015). *Fire Service Culture - Firefighter Nation: Fire Rescue - Firefighting News and Community.* [online] Firefighternation.com. Available at: https://www.firefighternation.com/firerescue/fire-service-culture/#gref [Accessed 11 Jul. 2022].

Firefighter Nation Staff (2016). *Poland's Big Investment - Firefighter Nation: Fire Rescue - Firefighting News and Community.* [online] Firefighternation.com. Available at: https://www.firefighternation.com/firerescue/polands-big-investment/#gref [Accessed 15 May 2019].

Firefighter Nation Staff (2017). *Challenging Tradition.* [online] Firefighter Nation: Fire Rescue - Firefighting News and Community. Available at: https://www.firefighternation.com/firerescue/challenging-tradition/#gref [Accessed 15 May 2019].

Fire Standards Board (2024). *Internal Governance and Assurance - Fire Standards Board %.* [online] Fire Standards Board. Available at: https://www.firestandards.org/standards/approved/internal-governance-and-assurance [Accessed 24 Apr. 2025].

Fletcher, B. (C) and Jones, F. (1992). Measuring Organizational Culture: The Cultural Audit. *Managerial Auditing Journal*, 7(6), pp.30–36. doi: https://doi.org/10.1108/eb017606.

Flood, J. (2010a). *The Fires.* New York: Penguin.

Flood, J. (2010b). *Why the Bronx Burned.* [online] New York Post. Available at: https://nypost.com/2010/05/16/why-the-bronx-burned/.

Fox, A. (2016). *How Louisville Metro Is Collaborating to Fight Fires.* [online] FireRescue1. Available at: https://www.firerescue1.com/communications-interoperability/articles/how-louisville-metro-is-collaborating-to-fight-fires-ezebXjlmIjGXvCy4/ [Accessed 15 May 2019].

Fresno Fire Department (2015). *Cortland Incident Multi -Agency Serious Accident Review Team.* [online] Available at: https://s3.documentcloud.org/documents/2694283/Sart.pdf [Accessed 10 Oct. 2021].

Fryer, K., Antony, J. and Ogden, S. (2009). Performance Management in the Public Sector. *International Journal of Public Sector Management*, 22(6), pp.478–498. doi: https://doi.org/10.1108/09513550910982850.

Galea, E. (2019). *Interim phase 1 recommendations for the Grenfell inquiry signature* [online] Available at: https://gala.gre.ac.uk/id/eprint/25970/1/25970%20GALEA_Interim_Recommendations_Grenfell_Inquiry_2019.pdf [Accessed 8 Oct. 2022].

Gagliano, M., Phillips, C., Jose, P. and Bernocco, S. (2008). *Air Management for the Fire Service*. Tulsa, Okla.: Pennwell.

Gasaway, R. (2018). *Explaining Resistance to Change - Situational Awareness Matters*. [online] Situational Awareness Matters. Available at: https://www.samatters.com/explaining-resistance-change/ [Accessed 10 Aug. 2022].

Gasaway, R.P. (2009). *Fireground Command Decision Making: Understanding the Barriers Challenging Command Situation Awareness*. Roseville, Mn: Gasaway Consulting Group Llc.

Gernay, T., Ni, S., Unobe, D., Lucherini, A., Chaudhary, R. and Van Coile, R. (2023). Cost–Benefit Analysis of Fire Protection in Buildings: Application of a Present Net Value Approach. *Fire Technology*, 59. doi: https://doi.org/10.1007/s10694-023-01419-2

Gianluca, P. and David, A. (2015). A definition of cascading disasters and cascading effects: Going beyond the 'toppling dominos' metaphor. *GRF Davos Planet@Risk*, 3(1), pp.58–67.

Gough, B. (2017) 'Fire Service Culture' in Grimwood, P. (2017). *Euro Firefighter 2: Firefighting Tactics and Fire engineer's Handbook: 6,701 Building Fires*. Huddersfield, West Yorkshire D&M Heritage Press. pp 442-446.

Gourbatsis, A. (2014). *Investigation into the Death of Mathaios Mantzios. Technical Report*. Andrianos Gourbatsis.

Gourbatsis, A. (2015). *The Cost of Forest Firefighting in Greece*. [online] Available at: https://www.eglimata-emprismou.gr/%CF%84%CE%BF-%CE%BA%CF%8C%CF%83%CF%84%CE%BF%CF%82-%CF%84%CE%B7%CF%82-%CE%B4%CE%B1%CF%83%CE%BF%CF%80%CF%85%CF%81%CF%8C%CF%83%CE%B2%CE%B5%CF%83%CE%B7%CF%82-%CF%83%CF%84%CE%B7%CE%BD-%CE%B5%CE%BB%CE%BB/ [Accessed 20 Apr. 2024].

Gousopoulou, S. and Hourdaki, K. (1994). *Η Πρόληψη Της Υγείας Και Της Ασφάλειας Του Προσωπικού Της Πυροσβεστικής Υπηρεσίας (Π.Υ.)*. [online] www.firesecurity.gr. Available at: https://www.firesecurity.gr/bibliothiki/kepe.htm [Accessed 15 May 2021].

Graham, F. (2013). Can Big Data Help Fight Fires and save lives? *BBC News*. [online] 22 Mar. Available at: https://www.bbc.com/news/business-21902070 [Accessed 3 Jul. 2018].

Graveling, R. and Crawford, J. (2010). *Occupational Health Risks in Firefighters*. [online] Available at: https://assets.publishing.service.gov.uk/media/5a7df54be5274a2e8ab44ebc/occu-health-risk-in-firefighters.pdf [Accessed 18 Jun. 2020].

Greek Government (1998). *Assignment of Forest Firefighting to the Hellenic Fire Service*. 2612/1998: FEK112/A/25.5.1998. Athens. National Printing Office.

Greek Government (1999) *Common Ministerial Decree 12030.F.109.1/1999 - FEK 713/B/19-5-1999. Measures to Ensure Cooperation and Coordination between the Fire service, the Military, the police, Local authorities, the Health Service and Other Agencies Assisting in Forest Fire Protection and Firefighting*. Athens. National Printing Office.

Greiving, S., Fleischhauer, M. and Wanczura, S. (2006). Management of natural hazards in Europe: The role of spatial planning in selected EU member states. *Journal of Environmental Planning and Management*, 49(5), pp.739–757. doi: https://doi.org/10.1080/09640560600850044.

Grenfell Tower Inquiry (2019). *Phase 1 Report | Grenfell Tower Inquiry*. [online] Grenfelltowerinquiry.org.uk. Available at: https://www.grenfelltowerinquiry.org.uk/phase-1-report [Accessed 4 Nov. 2022].

Grierson, J. and Kirk, A. (2022). Firefighter Numbers in England down 20% since 2010, Analysis Shows. *The Guardian.* [online] 13 Aug. Available at: https://www.theguardian.com/uk-news/2022/aug/13/firefighter-numbers-in-england-down-20-since-2010-analysis-shows [Accessed 13 Feb. 2024].

Grimwood, P. (2006). *SOG4242/2006/7 Snatch Rescues.* [online] Available at: chrome-extension://efaidnbmnnnibpcajpcglclefindmkaj/http://www.cfbt-be.com/images/teksten/Grimwoods-Snatch-Rescues.pdf [Accessed 24 Mar. 2020].

Grimwood, P. (2008). *Euro firefighter: Global Firefighting Strategy and tactics.* Huddersfield, England: Jeremy Mills Publishing.

Grimwood, P. (2009). *Model SOP Standard Operating Procedure: Tactical Deployment into Fire Involved Structures, Version 2.* [online] Available at: www.firetactics.com [Accessed 18 Jun. 2020].

Grimwood, P. (2017). *Euro Firefighter 2: Firefighting Tactics and Fire engineer's Handbook: 6,701 Building Fires.* Huddersfield, West Yorkshire D&M Heritage Press.

Guidotti, T.L. and Clough, V.M. (1992). Occupational Health Concerns of Firefighting. *Annual Review of Public Health*, 13(1), pp.151–171. doi:https://doi.org/10.1146/annurev.pu.13.050192.001055.

Hammar, C., Høgskulen På Vestlandet, W., Pareto, L., Heldal, I. and Wijkmark, C. (2016). *Simulation and Serious Games for Firefighter training: Challenges for Effective Use Simulation and Serious Games for Firefighter training: Challenges for Effective use.* [online] Available at: https://www.researchgate.net/publication/311442216_SIMULATION_AND_SERIOUS_GAMES_FOR_FIREFIGHTER_TRAINING_CHALLENGES_FOR_EFFECTIVE_USE [Accessed 2 Aug. 2019].

Hampshire Fire and Rescue Service (2013). *Report of the Hampshire Fire and Rescue Service Investigation into the Deaths of Firefighters Alan Bannon and James Shears in Flat 72, Shirley Towers, Church Street, Southampton, SO15 5PE, on Tuesday 6 April 2010.* Hampshire: Hampshire Fire and Rescue Service.

Harrami, O. and McIntyre, C. (2006). *Fire and Fire Protection in Homes and Public Buildings an Analysis of Swedish Fire Statistics and Fire Protection Strategies. Report Number 1/06*. Stockholm: Swedish Chemicals Inspectorate.

Hartin, E. (2007). *Case Studies in Extreme Fire Behaviour: 12 Minutes on the Fireground.* [online] Building a Better Fire Department FRI 2007. Available at: https://www.cfbt-us.com [Accessed 4 Nov. 2018].

Hasenmeier, P. (2008). *The History of Firefighter Personal Protective Equipment*. [online] Fire Engineering. Available at: https://www.fireengineering.com/fire-prevention-protection/the-history-of-firefighter-personal-protective-equipment/#gref [Accessed 27 Jan. 2020].

Health and Safety Executive (2010). *Striking the Balance between Operational and Health and Safety Duties in the Fire and Rescue Service*. Caerphilly: Health and Safety Executive.

Health and Safety Executive (2011). *HSE - Heroism in the Fire and Rescue Service*. [online] www.hse.gov.uk. Available at: https://www.hse.gov.uk/services/fire/heroism.htm.

Hellenic Fire Service (1992). *Internal Fire Service Policy: Presidential Order 210/16.6.1992 (FEK - 99A)*. Athens: National Government.

Henderson, M., Kostas Kalabokidis, Marmaras, E., Konstantinidis, P. and Manussos Marangudakis (2005). Fire and Society: A Comparative Analysis of Wildfire in Greece and the United States. *Human Ecology Review*, [online] 12(2), pp.169–182. Available at: https://www.researchgate.net/publication/258769431.

Heydari, A., Ostadtaghizadeh, A., Ardalan, A., Ebadi, A., Mohammadfam, I. and Khorasani-Zavareh, D. (2021). Exploring the Criteria and Factors Affecting Firefighters' resilience: a Qualitative Study. *Chinese Journal of Traumatology*, 25(2). doi: https://doi.org/10.1016/j.cjtee.2021.06.001.

His Majesty's Inspectorate of Constabularies and Fire and Rescue Services (2018). *Fire and Rescue Service Inspections 2018/19 Summary of Findings from Tranche 1.* Birmingham: HMICFRS.

His Majesty's Inspectorate of Constabularies and Fire and Rescue Services (2023). *Values and Culture in Fire and Rescue Services.* Birmingham: HMICFRS.

His Majesty's Inspectorate of Constabulary and Fire and Rescue Services (2017). *Proposed Fire and Rescue Services Inspection Programme and Framework 2018/19.* Birmingham: HMICFRS.

His Majesty's Inspectorate of Constabulary and Fire and Rescue Services (2019). *Too Much Variation in Fire and Rescue services, Says Report - His Majesty's Inspectorate of Constabulary and Fire & Rescue Services.* [online] His Majesty's Inspectorate of Constabulary and Fire & Rescue Services. Available at: https://hmicfrs.justiceinspectorates.gov.uk/news/news-feed/too-much-variation-in-fire-and-rescue-services-says-report/ [Accessed 10 Aug. 2022].

HM Government (2008). *Fire Service Manual, Volume 2. Fire Service Operations: Incident Command (3rd ed).* Norwich: The Stationery Office.

Hocy, I. (2023). *Firefighters Have the Most Dangerous Job in the UK.* [online] International Fire & Safety Journal. Available at: https://internationalfireandsafetyjournal.com/firefighters-have-the-most-dangerous-job-in-the-uk/#:~:text=A%20new%20study%20by%20Forbes [Accessed 13 Aug. 2023].

Home Office (2016). *Fire and Rescue Authorities: Operational Statistics Bulletin for England 2015 to 2016.* London: Home Office.

Home Office (2025). *Fire and Rescue Incident Statistics: England, Year Ending March 2025.* [online] London: Home Office. Available at: https://www.gov.uk/government/statistics/fire-and-rescue-incident-statistics-year-ending-march-2025/fire-and-rescue-incident-statistics-year-ending-march-2025 [Accessed 30 March 2025].

Home Office (2023). *Fire Statistics Data Tables*. [online] www.gov.uk. Available at: https://www.gov.uk/government/statistical-data-sets/fire-statistics-data-tables#workforce-and-workforce-diversity [Accessed 10 Dec. 2023].

Hossain, N., Fillaili, R., Lubaale, G., Mulumbi, M. and Mariz Tadros, M.R. (2010). *The Social Impacts of Crisis: Findings from Community-Level Research in Five Developing Countries. SSRN Electronic Journal*. doi:https://doi.org/10.2139/ssrn.1879128.

https://www.youtube.com/@Halifax Regional Municipality (2016). *Incident Command System Scenario Training*. [online] www.youtube.com. Available at: https://www.youtube.com/watch?v=BNI5EFMCwZ8 [Accessed 15 May 2019].

Hytten, K. and Hasle, A. (1989). Fire fighters: a Study of Stress and Coping. *Acta Psychiatrica Scandinavica*, 80(s355), pp.50–55. doi:https://doi.org/10.1111/j.1600-0447.1989.tb05253.x.

Jick, T. and Peiperl, M. (2011). *Managing Change: Cases and Concepts*. Boston: Mcgraw-Hill/Irwin.

Inman, P. (2017). *A Decade of overspending: How Greece Plunged into Economic Crisis*. [online] The Guardian. Available at: https://www.theguardian.com/world/2015/jul/03/greece-overspending-defence-wages-taxation-economic-crisis.

International Association of Fire Chiefs (2021). *Pre-Incident*. [online] Available at: https://www.iafc.org/docs/default-source/1haz/respondingtoelectricalvehiclefires.pdf?sfvrsn=9421650c_6 [Accessed 21 Jul. 2022].

Institution of Fire Engineers (2023). *IFE Reveals New Firefighter Safety Database at Centenary Launch*. [online] Ife.org.uk. Available at: https://www.ife.org.uk/IFE-News/ife-reveals-new-firefighter-safety-database-at-centenary-launch [Accessed 13 Apr. 2019].

International Fire Service Journal of Leadership and Management (2013). *International Fire Service Journal of Leadership and Management*. [online] Available at: https://www.ifsjlm.org/sites/default/files/past-edition-pdfs/IFSJLM_Vol7.pdf [Accessed 10 Oct. 2021].

International Fire Service Training Association (2018) *Essentials of Fire Fighting and Fire Department Operations*. 7th ed. Stillwater, OK: International Fire Service Training Association.

Isenson, N. (2019). *Aging Strikes Germany's Firefighters*. [online] dw.com. Available at: https://www.dw.com/en/germanys-firefighters-on-the-decline-as-aging-strikes/a-49842885 [Accessed 10 Aug. 2022].

ITVX News (2019). *Union Slams New 'rapid response' Fire Vehicles*. [online] ITVX News. Available at: https://www.itv.com/news/border/2019-03-28/union-slams-new-rapid-response-fire-vehicles [Accessed 11 Apr. 2021].

Jiang, Y. and Yuan, Y. (2019). Emergency Logistics in a Large-Scale Disaster Context: Achievements and Challenges. *International Journal of Environmental Research and Public Health*, 16(5), p.779. doi: https://doi.org/10.3390/ijerph16050779.

Johansson, R. (2007). On Case Study Methodology. *Open House International*, 32(3), pp.48–54. doi: https://doi.org/10.1108/ohi-03-2007-b0006.

K. Bradley Penuel and Statler, M. (2010). *Encyclopedia of Disaster Relief*. New York: SAGE Publications.

Kafantari, H. (2014). *The Tetra Communications System and the Hellenic Fire Service*. [online] https://kafantari.com/. Available at: https://kafantari.com/2014/10/30/%cf%83%cf%85%cf%83%cf%84%ce%ae%ce%bc%ce%b1%cf%84%ce%b1-%ce%b5%cf%80%ce%b9%ce%ba%ce%bf%ce%b9%ce%bd%cf%89%ce%bd%ce%af%ce%b1%cf%82-%cf%84%cf%8d%cf%80%ce%bf%cf%85-tetra-%cf%83%cf%84%ce%bf-%cf%80%cf%85/#more-1412 [Accessed 18 Jun. 2021].

Kalyvas, S.N. (2015). *Modern Greece: What Everyone Needs to Know*. Oxford: Oxford University Press.

Kanchana H. (2024). *Health Safety FRS Lessons Learned.* [online] Scribd. Available at: https://www.scribd.com/document/685340539/Health-Safety-FRS-Lessons-Learned [Accessed 9 Aug. 2020].

Kanterman, R.E. (2014). *Courage to Be Safe/ Everyone Goes Home: a Look inside the Program*. [online] *www.fireengineering.com*. Available at: https://www.fireengineering.com/firefighting/feu-courage-to-be-safe-everyone-goes-home-a-look-inside-the-program/#gref [Accessed 6 Nov. 2018].

Karlsson, B. and Svensson, S. (n.d.). A Decade of Scandinavian Research Aimed at Benefitting the Fire Service. [online] Available at: chrome-extension://efaidnbmnnnibpcajpcglclefindmkaj/https://fireseat.eng.ed.ac.uk/sites/fireseat.eng.ed.ac.uk/files/images/bjorn_paper.pdf [Accessed 6 Nov. 2018].

Kastanakis, M.N. and Voyer, B.G. (2014). The Effect of Culture on Perception and cognition: a Conceptual Framework. *Journal of Business Research*, [online] 67(4), pp.425–433. doi: https://doi.org/10.1016/j.jbusres.2013.03.028.

Katsabas T. (2016). *Five Myths and Five Truths about the Greek Financial Crisis*. [online] Kathimerini. Available at: https://www.kathimerini.gr/opinion/852855/pente-mythoi-kai-pente-alitheies-gia-tin-elliniki-oikonomiki-krisi/ [Accessed 11 Jul. 2022].

Kentikelenis, A., Karanikolos, M., Papanicolas, I., Basu, S., McKee, M. and Stuckler, D. (2011). Health Effects of Financial crisis: Omens of a Greek Tragedy. *The Lancet*, 378(9801), pp.1457–1458. doi: https://doi.org/10.1016/s0140-6736(11)61556-0.

Kerber, S. (2011). Analysis of Changing Residential Fire Dynamics and Its Implications on Firefighter Operational Timeframes. *Fire Technology*, 48(4), pp.865–891. doi: https://doi.org/10.1007/s10694-011-0249-2.

Killen, B., Stittleburg, P.C., Webb, W.M. and Siarnicki, R.J. (2005). *Radio Communications for the Fire Service: a Planning Guide for Obtaining the Communications System You Need for Enhanced Safety and Emergency Preparedness*. USA: IAFC.

Kilmann, R.H., Saxton, M.J. and Serpa, R. (1986). Issues in Understanding and Changing Culture. *California Management Review*, 28(2), pp.87–94. doi: https://doi.org/10.2307/41165186.

Kindreich, A. (2017). *The Greek Financial Crisis (2009–2016) - Financial Scandals, Scoundrels & Crises*. [online] Financial Scandals, Scoundrels & Crises. Available at: https://www.econcrises.org/2017/07/20/the-greek-financial-crisis-2009-2016/ [Accessed 12 Jan. 2020].

Klaene, B.J. (2015). *Structural Firefighting*. Massachusetts: Jones & Bartlett Publishers.

Kloot, L. (2009). Performance Measurement and Accountability in an Australian Fire Service. *International Journal of Public Sector Management*, 22(2), pp.128–145. doi: https://doi.org/10.1108/09513550910934538.

Kouras, B. (2018). *The Greek Diaspora Approaches the 5 million Mark*. [online] Greek City Times. Available at: https://greekcitytimes.com/2018/11/18/the-greek-diaspora-approaches-the-5-million-mark/?amp [Accessed 15 May 2019].

Koutsoukis, K.S. (2003). Political Corruption in Greece. *Palgrave Macmillan UK eBooks*, pp.24–36. doi: https://doi.org/10.1057/9781403919991_3.

Kunadharaju, K., Smith, T.D. and DeJoy, D.M. (2011). Line-of-duty Deaths among U.S. firefighters: an Analysis of Fatality Investigations. *Accident Analysis & Prevention*, 43(3), pp.1171–1180. doi: https://doi.org/10.1016/j.aap.2010.12.030.

Kunavut, K. (2014). An Overview of Digital Trunked Radio: Technologies and Standards. *The Journal of Industrial Technology*, [online] 10(2). Available at: https://www.researchgate.net/publication/265908152 [Accessed 6 Nov. 2018].

Lamb, K.J., Davies, J., Bowley, R. and Williams, J.-P. (2014). Incident Command training: the Introspect Model. *International Journal of Emergency Services*, 3(2), pp.131–143. doi: https://doi.org/10.1108/ijes-09-2013-0023.

Lambert, K. (2015). *The Smoke Stopper*. [online] Available at: http://www.cfbt-be.com/images/artikelen/artikel_27_EN.pdf [Accessed 29 Apr. 2019].

Lambert K. (2014). *Learning from Incidents*. [online] Cfbt-be.com. Available at: http://www.cfbt-be.com/nl/leren-uit-incidenten [Accessed 12 Sep. 2022].

Lambert, K. (2012). Firefighting: Command and Tactics. [online] Available at: chrome-extension://efaidnbmnnnibpcajpcglclefindmkaj/http://www.cfbt-be.com/images/artikelen/artikel_11_ENG.pdf [Accessed 6 Nov. 2018].

Lambert, K. and Baaij, S. (2015). *Fire Dynamics. Technical Approach, Tactical Application*. The Hague: SDU Publications.

Lambropoulos, V. (2020). *An Age of Crime. Corrputpion in the Greek Fire Service*. [online] To Vima. Available at: https://www.tovima.gr/2020/08/08/society/egklima-diarkeias-i-diafthora-stin-pyrosvestiki/ [Accessed 23 Jun. 2022].

Lambropoulou, E. (2012). Myths and Realities about Corruption in Public Administration and its Discourse in Greece. *Amsterdam Law Forum*, 4(3), p.77. doi: https://doi.org/10.37974/alf.231.

Lane, B. (2018). *Communicating with the Residents in an Emergency REPORT OF on Behalf of on Instructions of Subject Matter Inspection Date(s)*. [online] Available at: https://assets.grenfelltowerinquiry.org.uk/documents/Dr%20Barbara%20Lane%20report%20-%20section%2018%20%28Phase%201%20-%20supplemental%29.pdf [Accessed 10 May 2020].

Launder, D. and Perry, C. (2014). A Study Identifying Factors Influencing Decision Making in Dynamic Emergencies like Urban Fire and Rescue Settings. *International Journal of Emergency Services*, 3(2), pp.144–161. doi: https://doi.org/10.1108/ijes-06-2013-

0016.

Lee, A. and Meyer, R. (2000). *NOVA Online | Escape! | Escape through Time: Fire*. [online] www.pbs.org. Available at: https://www.pbs.org/wgbh/nova/escape/timefire.html [Accessed 12 Jan. 2020].

Local Government Association (2017). *Fire and Rescue Services in England: a Guide for Police and Crime Panel Members*. London: Local Government Association.

Local Government Association (2017). *An Overview of the English Fire and Rescue Service. Police and Crime Panel Workshop*. London: Home Office.

London Fire Brigade (2024). *Firefighting in high-rise Buildings*. [online] London-fire.gov.uk. Available at: https://www.london-fire.gov.uk/about-us/services-and-facilities/techniques-and-procedures/firefighting-in-high-rise-buildings [Accessed 29 Apr. 2024].

London Fire Brigade (2023). *'Brigade Will Remain World Class and Londoners Will Still Be safe,' says Authority chief, Ahead of Station Closures*. [online] London-fire.gov.uk. Available at: https://www.london-fire.gov.uk/news/2014-news/brigade-will-remain-world-class-and-londoners-will-still-be-safe-says-authority-chief-ahead-of-station-closures/ [Accessed 14 Aug. 2024].

London Fire Brigade (2020). *Improving the Operational Safety Management Model. Policy Number LFC-0374*. London: London Fire Brigade.

London Fire Brigade (2017). *Fire Facts Fires in Greater London 1966 - 2016*. London: London Fire Brigade.

London Fire Brigade (2015). *Fire Facts Incident Response Times*. London: London Fire Brigade.

London Fire Brigade (2013). *Getting to Emergency Incidents as Quickly as Possible Fifth London Safety Plan Supporting Document No. 8 Consultation Draft*. London: London Fire Brigade.

London Fire and Emergency Planning (2013). *Coroner's Inquests following the fire Lakanal House on 3 July 2009*. London. LFEPA

Lupyan, G. (2017). How Reliable Is Perception? *Philosophical Topics*, 45(1), pp.81–106. doi: https://doi.org/10.5840/philtopics201745115.

Madrzykowski, D. (2010). *Understanding and implementing the 16 Life Safety Initiatives – Chapter 3: Research and Data Collection in the Fire Service*. Fire Protection Publications, Stillwate

Maher, K., Bateman, N. and Randall, R. (2019). Fire and Rescue Operational effectiveness: the Effect of Alternative Crewing Patterns. *Production Planning & Control*, 31, pp.1–12. doi: https://doi.org/10.1080/09537287.2019.1701232.

Manning, L. (2019). Grenfell Tower fire: 'Systemic failures' in Fire brigade's Response. *BBC News*. [online] 29 Oct. Available at: https://www.bbc.com/news/uk-50216606 [Accessed 15 May 2020].

Manolopoulos, D. (2017). An Evaluation of Employee Motivation. *Employee Relations*, 30(1), pp.63–85. doi: https://doi.org/10.1108/01425450810835428.

March, J. G. (1991). Exploration and Exploitation in Organizational Learning. *Organization Science*, *2*(1), 71–87. http://www.jstor.org/stable/2634940

Markley, R. (2016). *The Fire Service Future Will Be data-driven.* [online] FireRescue1. Available at: https://www.firerescue1.com/firefighter-training/articles/the-fire-service-future-will-be-data-driven-SVuImbg7dB70rOdH/ [Accessed 11 Aug. 2024].

Markovits, Y., Davis, A.J. and van Dick, R. (2007). Organizational Commitment Profiles and Job Satisfaction among Greek Private and Public Sector Employees. *International Journal of Cross-Cultural Management*, 7(1), pp.77–99. doi: https://doi.org/10.1177/1470595807075180.

Marr, B. (2017). *Is Big Data Analytics the Secret to Successful Fire Fighting?* [online] Forbes. Available at: https://www.forbes.com/sites/bernardmarr/2017/01/13/is-big-data-analytics-the-secret-to-successful-fire-fighting/ [Accessed 15 May 2019].

McCormack, C. (2022) *Fire Department Strategic Planning: Informed Fireground Leadership. Fire Engineering.* Available at: https://www.fireengineering.com/firefighting/fire-department-strategic-planning-informed-fireground-leadership/ (Accessed: 24 April 2024).

McDonough, J. (2018). *The Science of Water Extinguishment - International Innovations - John McDonough.* [online] Firehouse. Available at: https://www.firehouse.com/operations-training/hoselines-water-appliances/article/12395735/the-science-of-water-extinguishment-international-innovations-john-mcdonough [Accessed 11 Apr. 2019].

McManis Associates and John T. O'Hagan and Associates (1984). *Dallas Fire Department Staffing Level Study*. Dallas: Dallas Fire Department.

Mercer, J., Gaillard, J.C., Crowley, K., Shannon, R., Alexander, B., Day, S. and Becker, J. (2012). Culture and Disaster Risk reduction: Lessons and Opportunities. *Environmental Hazards*, 11(2), pp.74–95. doi: https://doi.org/10.1080/17477891.2011.609876.

Metz, J. (2015). *Can Leadership Be Taught – a Study about Leadership Development in Education.* Master's Thesis. Sweden: Linnaeus University.

Michaelides, M.A., Parpa, K.M., Henry, L.J., Thompson, G.B. and Brown, B.S. (2011). Assessment of Physical Fitness Aspects and Their Relationship to Firefighters Job Abilities. *Journal of Strength and Conditioning Research*, 25(4), pp.956–965. doi: https://doi.org/10.1519/jsc.0b013e3181cc23ea.

Moore-Merrell, L., McDonald, S., Zhou, A., Fisher, E. and Moore, J. (2006). *Contributing Factors to Line of Duty Deaths*. USA: International Association of Fire Fighters.

Mora, W.R. (2016). *Preventing Firefighter Disorientation.* [online] Google Books. Available at: https://books.google.gr/books?hl=en&lr=&id=YwlpDwAAQBAJ&oi=fnd&pg=PR7&dq=firefighter+disorientation&ots=ueU_PKlKEH&sig=_xQiA3PwW-nm5JQg5mDZIzJFycA&redir_esc=y#v=onepage&q=firefighter%20disorientation&f=false [Accessed 10 Aug. 2021].

Murphy, A. (2014). *The Public's Perception of the Modern-Day Fire and Rescue Service.* Undergraduate Dissertation. Lancaster University.

Murphy, B.L. (2007). Locating Social Capital in Resilient community-level Emergency Management. *Natural Hazards*, 41(2), pp.297–315. doi: https://doi.org/10.1007/s11069-006-9037-6.

Murphy, J.K. (n.d.). *NFPA 1001 Standards as Minimum Training Standard for Firefighters - HGExperts.com.* [online] www.hgexperts.com. Available at: https://www.hgexperts.com/expert-witness-articles/nfpa-1001-standards-as-minimum-training-standard-for-firefighters-19151 [Accessed 16 Jul. 2022].

Murray, N. (2024) 'We need to talk about learning lessons', *Fire Magazine*. Available at: https://fire-magazine.co.uk/we-need-to-talk-about-learning-lessons (Accessed: 24 April 2025).

Napp, C. (2015). *Identifying Barries for change. What Is Preventing the Acceptance of the New Fire Attack Strategy?* [online] U.S. Fire Administration. Available at: https://apps.usfa.fema.gov/pdf/efop/efo49561.pdfhttps://apps.usfa.fema.gov/pdf/efop/efo49561.pdf [Accessed 20 Apr. 2023].

Narens, L. (2003). A theory of belief. *Journal of Mathematical Psychology*, 47(1), pp.1–31. doi: https://doi.org/10.1016/s0022-2496(02)00017-2.

National Audit Office (NAO). (2015). *Impact of funding reductions on fire and rescue services - NAO report.* [online] Available at: https://www.nao.org.uk/reports/impact-of-funding-reductions-on-fire-and-rescue-services [Accessed 20 Apr. 2022].

National Fire Chiefs Council (2023) *Operational Response Implementation Guide*. National Fire Chiefs Council. Available at: https://nfcc.org.uk/our-services/national-operational-guidance/tools/the-operational-response-implementation-guide/ (Accessed: 24 April 2025).

National Fire Chiefs' Association (n.d.). *National Operational Learning Good Practice Guide for Fire and Rescue Services*. London: National Fire Chiefs Association.

National Fire Chiefs Council (n.d.). *Command Roles and Responsibilities.* [online] NFCC. Available at: https://nfcc.org.uk/control-measure/control-measure-command-roles-and-responsibilities/ [Accessed 15 May 2020].

National Fire Protection Association (2024). *Fireworks | NFPA*. [online] www.nfpa.org. Available at: https://www.nfpa.org/education-and-research/home-fire-safety/fireworks [Accessed 14 Aug. 2024].

National Fire Protection Association (2015). Data in the Fire Service: Can We Improve Our Planning and Response While Documenting Our Value to Our Community? Chicago. NFPA.

National Institute for Standards and Technology (2010). *Landmark Residential Fire Study Shows How Crew Sizes and Arrival Times Influence Saving Lives and Property*. [online] https://www.nist.gov/. Available at: https://www.nist.gov/news-events/news/2010/04/landmark-residential-fire-study-shows-how-crew-sizes-and-arrival-times [Accessed 10 Oct. 2021].

National Institute for Standards and Technology (2024). Wear and Tear May Cause Firefighter Gear to Release More 'Forever Chemicals'. *NIST*. [online] Available at: https://www.nist.gov/news-events/news/2024/01/wear-and-tear-may-cause-firefighter-gear-release-more-forever-chemicals#:~:text=Last%20year%2C%20a%20study%20by [Accessed 25 Mar. 2024].

NBC Universal (2010). *No pay, no spray: Firefighters Let Home Burn.* [online] NBC News. Available at: https://www.nbcnews.com/id/wbna39516346#.Won9zVu0PIV [Accessed 10 Aug. 2022].

New South Wales Rural Fire Service (2017). *Fast Facts - NSW Rural Fire Service.* [online] Nsw.gov.au. Available at: https://www.rfs.nsw.gov.au/about-us/fast-facts [Accessed 15 May 2019].

News Beast (2018). *The Fire on the Ship Eleftherios Venizelos Is Extinguished.* [online] Newsbeast. Available at: https://www.newsbeast.gr/greece/arthro/3945311/esvise-i-fotia-sto-ploio-eleytherios-venizelos [Accessed 11 Jul. 2022].

Newsroom (2011). *Προκήρυξη Πυροσβεστικής Για 4000 Θέσεις 2011.* [online] Dikaiologitika.gr. Available at: https://www.dikaiologitika.gr/arxeio/957/prokiryksi-pyrosvestikis-gia-4000-theseis-2011 [Accessed 15 May. 2020].

Office of the Deputy Prime Minister (2006). *The Economic Cost of Fire: Estimates for 2004.* West Yorkshire: ODPM Publications.

Office of the Fire Marshal (2002). *OPERATE COMMUNICATION EQUIPMENT We acknowledge the input of the Standard Review Advisory Committee.* [online] Available at: https://www2.gnb.ca/content/dam/gnb/Departments/ps-sp/pdf/Safety_Protection/training/commeq_e.pdf [Accessed 4 Nov. 2018].

Organisation for Economic Cooperation and Development (2024). *Taming Wildfires in the Context of Climate change: the Case of Greece OECD ENVIRONMENT POLICY PAPER NO. 43 COUNTRY CASE STUDY.* [online] Available at: https://www.oecd.org/content/dam/oecd/en/publications/reports/2024/06/taming-wildfires-in-the-context-of-climate-change-the-case-of-greece_754f2c60/cfb797a7-en.pdf [Accessed 7 Dec. 2024].

Orthodox Church in America (2024). *The Orthodox Faith - Volume III - Church History - Sixteenth Century - the Greek Orthodox under the Ottoman Turks*. [online] Oca.org. Available at: https://www.oca.org/orthodoxy/the-orthodox-faith/church-history/sixteenth-century/the-greek-orthodox-under-the-ottoman-turks [Accessed 11 Apr. 2021].

Otker-Robe, I. and Podpiera, A.M. (2013). *The Social Impact of Financial Crises: Evidence from the Global Financial Crisis*. [online] *papers.ssrn.com*. Rochester, NY: The World Bank. Available at: https://ssrn.com/abstract=2354754 [Accessed 6 Nov. 2018].

Ozturk, S. and Sozdemir, A. (2015). Effects of Global Financial Crisis on Greece Economy. *Procedia Economics and Finance*, [online] 23(23), pp.568–575. doi: https://doi.org/10.1016/s2212-5671(15)00441-4.

Pan-Hellenic Contract Firefighters Union (2017). *Allocation of Hellenic Fire Service Budget 2017*. [online] https://www.popysyp.gr/. Available at: https://www.popysyp.gr/ νομοθεσια/νόμοι/377-κατανομή-πιστώσεων-προϋπολογισμού-πυροσβεστικού-σώματος-έτους-2017.html [Accessed 14 Aug. 2024].

Pantelides, K. and Tziritis, E. (2009). *Volunteers, Forest Fire Protection and Forest Firefighting: A Proposal for the Effective Inlcusion of Volunteers in Forest Protection*. Athens: WWF Greece.

Papaconstantinou, P., Tsagkanos, A.G. and Siriopoulos, C. (2013). How bureaucracy and corruption affect economic growth and convergence in the European Union. *Managerial Finance*, 39(9), pp.837–847. doi: https://doi.org/10.1108/mf-12-2009-0143.

Papaioannou, S. (2021). *The History - and Future - of Greek Firefighting* - [online] GreekReporter.com. Available at: https://greekreporter.com/2021/08/27/the-history-and-future-of-greek-firefighting/ [Accessed 25 Apr. 2024].

Pappas, T.S. (2010). *The Causes of the Greek Crisis Are in Greek Politics*. [online] OpenDemocracy. Available at: https://www.opendemocracy.net/en/openeconomy/causes-of-greek-crisis-are-in-greek-politics/ [Accessed 27 Jan. 2020].

Pennington, M.L., Cardenas, M., Nesbitt, K., Coe, E., Kimbrel, N.A., Zimering, R.T. and Gulliver, S.B. (2021). Career versus Volunteer firefighters: Differences in Perceived Availability and Barriers to Behavioral Health care. *Psychological Services*, 19(3). doi: https://doi.org/10.1037/ser0000559.

Pescaroli, G. and Alexander, D. (2016). Critical infrastructure, panarchies and the vulnerability paths of cascading disasters. *Natural Hazards*, 82(1), pp.175–192. doi: https://doi.org/10.1007/s11069-016-2186-3.

Petmezas, S. (2024). *History of Modern Greece.* [online] Scribd. Available at: https://www.scribd.com/document/129416168/History-of-modern-Greece-by-S-Petmezas [Accessed 10 Aug. 2021].

Pidd, H. (2011). German stereotypes: über-efficiency. *The Guardian.* [online] 18 Mar. Available at: https://www.theguardian.com/world/2011/mar/18/german-stereotypes-uber-efficiency [Accessed 27 Jan. 2020].

Pillai, R. and Williams, E.A. (2004). Transformational leadership, self-efficacy, Group cohesiveness, commitment, and Performance. *Journal of Organizational Change Management*, 17(2), pp.144–159. doi: https://doi.org/10.1108/09534810410530584.

Pollack, K.M., Poplin, G.S., Griffin, S., Peate, W., Nash, V., Nied, E., Gulotta, J. and Burgess, J.L. (2017). Implementing Risk Management to Reduce Injuries in the U.S. Fire Service. *Journal of Safety Research*, 60, pp.21–27. doi: https://doi.org/10.1016/j.jsr.2016.11.003.

Pugno, M. (2009). Job Performance and Job Satisfaction: An Integrated Survey. *SSRN Electronic Journal*. doi: https://doi.org/10.2139/ssrn.1402566.

Rachman, M.M. (2021). The Impact of Work Stress and the Work Environment in the Organization: How Job Satisfaction Affects Employee Performance. *Journal of Human Resource and Sustainability Studies*, 09(02), pp.339–354. doi: https://doi.org/10.4236/jhrss.2021.92021.

Ragins, B.R. (1997). Diversified Mentoring Relationships in Organizations: a Power Perspective. *The Academy of Management Review*, 22(2), pp.482–521.

Ramesh, M. (2009). Economic Crisis and Its Social Impacts. *Global Social Policy: An Interdisciplinary Journal of Public Policy and Social Development*, 9(1_suppl), pp.79–99. doi: https://doi.org/10.1177/1468018109106886.

Rapidsos, W. and Com (2015). *Outcomes Quantifying the Impact of Emergency Response Times*. [online] Available at: https://cdn2.hubspot.net/hubfs/549701/Documents/RapidSOS_Outcomes_White_Paper_-_2015_4.pdf [Accessed 18 Jun. 2020].

Rapp, C.E. and Wilson, R.S. (2022). Factors that contribute to trustworthiness across levels of authority in wildland fire incident management teams. *International Journal of Disaster Risk Reduction*, 73(73), p.102877. doi: https://doi.org/10.1016/j.ijdrr.2022.102877.

Reason, J. (1990). *Human Error*. Cambridge: Cambridge University Press.

Reason, J. (2000). Human Error: Models and Management. *British Medical Journal*, [online] 320(7237), pp.768–770. doi: https://doi.org/10.1136/bmj.320.7237.768.

Rielage, R. (2014). *5 Ways to Show Effects of Fire Budget Cuts*. [online] FireRescue1. Available at: https://www.firerescue1.com/fire-chief/articles/5-ways-to-show-effects-of-fire-budget-cuts-QFWFMFIp0xGdFUeu/.

Rinny, P., Bohlen Purba, C. and Handiman, U. (2020). The Influence of Compensation, Job Promotion, And Job Satisfaction on Employee Performance of Mercubuana University the Influence of Compensation, Job Promotion, And Job Satisfaction on Employee Performance of Mercubuana University. *www.ijbmm.com International Journal of Business Marketing and Management*, 5(2), pp.2456–4559. doi: https://www.researchgate.net/publication/362488344.

Ritchie, H. and Rosado, P. (2022). *Natural Disasters*. [online] Our World in Data. Available at: https://ourworldindata.org/natural-disasters [Accessed 20 Dec. 2022].

Rizospastis (2018). *Σοβαροί Κίνδυνοι Για Την Υγεία Και Την Ασφάλεια Των Πυροσβεστών | ΡΕΠΟΡΤΑΖ | ΡΙΖΟΣΠΑΣΤΗΣ*. [online] ΡΙΖΟΣΠΑΣΤΗΣ. Available at: https://www.rizospastis.gr/story.do?id=9775895 [Accessed 13 Aug. 2024].

Rogers, F.B., Rittenhouse, K.J. and Gross, B.W. (2015). The Golden Hour in trauma: Dogma or Medical folklore? *Injury*, [online] 46(4), pp.525–527. doi: https://doi.org/10.1016/j.injury.2014.08.043.

Roguski, J. and Szułczyńska, D. (2021). The Use of Innovative Technologies in Rescue and Firefighting Activities of the State Fire Service in Order to Increase the Safety of Officers and Protect the Environment. *Safety & Fire Technology*, 57(1), pp.100–113. doi: https://doi.org/10.12845/sft.57.1.2021.7.

Roman, J. (2022). *The PFAS Problem*. [online] Nfpa.org. Available at: https://www.nfpa.org/news-blogs-and-articles/nfpa-journal/2022/07/22/the-new-foam/foam-sidebar [Accessed 13 May 2024].

Sabetti, F. and Catanzaro, R. (1990). *Italian Politics, a Review: A Publication of the Conference Group on Italian Politics and the Carlo Cattaneo Institute*. [online] *Google Books*. F. Pinter. Available at: https://books.google.gr/books/about/Italian_Politics_a_Review.html?id=eLhDAQAAIAAJ&redir_esc=y [Accessed 4 Jun. 2018].

Sample, M., Thode, A.E., Peterson, C., Gallagher, M.R., Flatley, W., Friggens, M., Evans, A., Loehman, R., Hedwall, S., Brandt, L., Janowiak, M. and Swanston, C. (2022). Adaptation Strategies and Approaches for Managing Fire in a Changing Climate. *Climate*, [online] 10(4), p.58. doi: https://doi.org/10.3390/cli10040058.

Sangasubana, N. (2014). How to Conduct Ethnographic Research. *The Qualitative Report*, 16(2). doi: https://doi.org/10.46743/2160-3715/2011.1071.

Scattergood, A. (2017). Cuts to the Fire Service Are Putting Lives at Risk | Andrew Scattergood. *The Guardian*. [online] 15 Jun. Available at: https://www.theguardian.com/commentisfree/2017/jun/15/cuts-fire-service-putting-lives-at-risk [Accessed 9 Nov. 2018].

Schein, E.H. (2004). *Organizational Culture and Leadership (Jossey-Bass Business & Management series)*. Jossey Bass Inc.

Scholz, M., Gordon, D., Ramirez, L., Sigg, S., Dyrks, T. and Beigl, M. (2013). A Concept for Support of Firefighter Frontline Communication. *Future Internet*, 5(2), pp.113–127. doi: https://doi.org/10.3390/fi5020113.

Shaw, J. (2011). *Seniority versus Experience - Fire Fighting in Canada.* [online] Fire Fighting in Canada. Available at: https://www.firefightingincanada.com/seniority-versus-experience-7918/ [Accessed 10 Aug. 2022].

Smith, V., Abrams, A., Brakhage, C. and International Fire Service Training Association (2015). *Fire and Emergency Services Orientation and Terminology*. Stillwater, Okla.: Fire Protection Publications, Oklahoma State University.

Soler, D.J. (2014). *3 Reasons for Firefighters to Use the Radio | Fire Fighter Toolbox*. [online] Fire Fighter Toolbox. Available at: https://firefightertoolbox.com/3-reasons-use-radio/ [Accessed 11 Jul. 2022].

Spinney, N. (2014). *The Team Approach to Road Traffic Collisions.* [online] https://iffmag.com/. Available at: https://iffmag.com/the-team-approach-to-road-traffic-collisions/ [Accessed 15 May 2019].

Spell, J. (2022). *Firefighter training and certifications: Understanding the many levels*. [online] FireRescue1. Available at: https://www.firerescue1.com/firefighter-training/articles/firefighter-training-and-certifications-understanding-the-many-levels-bEikHYpeVviOiKWy/ [Accessed 20 Dec. 2022].

Stec, A.A., Dickens, K.E., Salden, M., Hewitt, F.E., Watts, D.P., Houldsworth, P.E. and Martin, F.L. (2018). Occupational Exposure to Polycyclic Aromatic Hydrocarbons and Elevated Cancer Incidence in Firefighters. *Scientific Reports*, [online] 8(1). doi: https://doi.org/10.1038/s41598-018-20616-6.

Stevenson, C. (2018). California wildfires: High Santa Ana Winds Bring Danger of Massive Blazes in South of State. *The Independent*. [online] 15 Oct. Available at: https://www.independent.co.uk/news/world/americas/california-wildfires-fire-wind-high-santa-ana-orange-county-storm-weather-a8584941.html [Accessed 14 Jul. 2019].

Stumpf, J. (1999). Incident Command System: the History and Need. the Internet Journal of Rescue and Disaster Medicine. *The Internet Journal of Rescue and Disaster Medicine*, 2(1).

Sukoco, B.M., Supriharyanti, E., Sabar, Susanto, E., Nasution, R.A. and Daryanto, A. (2021). Organisational Change Capacity and performance: the Moderating Effect of Coercive Pressure. *Asia-Pacific Journal of Business Administration*, 14(1), pp.27–49. doi: https://doi.org/10.1108/apjba-11-2020-0428.

Sun, Z., Strang, K. and Li, R. (2018). *(PDF) 10 Bigs: Big Data and Its Ten Big Characteristics*. [online] ResearchGate. Available at: https://www.researchgate.net/publication/322592851_10_Bigs_Big_Data_and_Its_Ten_Big_Characteristics.

Svensson, S. and Van de Veire, M. (2018). Experimental Study of Gas Cooling during Firefighting Operations. *Fire Technology*, 55(1), pp.285–305. doi: https://doi.org/10.1007/s10694-018-0790-3.

TA NEA (2018). *Αλέξανδρος Γρηγορόπουλος: Η Δολοφονία Που Άλλαξε Την Ελλάδα*. [online] TA NEA. Available at: https://www.tanea.gr/2018/12/06/greece/aleksandros-grigoropoulos-i-dolofonia-pou-allakse-tin-ellada/ [Accessed 14 Aug. 2019].

Taleb, N.N. (2007). The Black Swan: The Impact of the Highly Improbable. *The New York Times*. [online] 22 Apr. Available at: https://www.nytimes.com/2007/04/22/books/chapters/0422-1st-tale.html [Accessed 12 Jan. 2020].

Tarí, J.J., Molina-Azorín, J.F. and Heras, I. (2012). Benefits of the ISO 9001 and ISO 14001 standards: a Literature Review. *Journal of Industrial Engineering and Management*, 5(2). doi: https://doi.org/10.3926/jiem.488.

Taylor, J.A., Davis, A.L., Barnes, B., Lacovara, A.V. and Patel, R. (2015). Injury Risks of EMS responders: Evidence from the National Fire Fighter Near-Miss Reporting System. *BMJ Open*, 5(6), pp. e007562–e007562. doi: https://doi.org/10.1136/bmjopen-2014-007562.

Texas Department of Insurance (2016). *Firefighter Fatality Investigations: Annual Report FY 2016*. Texas: Texas Department of Insurance.

Thiel, A. and Stambaugh, H. (2013). *Improving Firefighter Communications: Special Report*. USA: Federal Emergency Management Agency.

Thompson, C. (2017). *Falling through Fire*. London: Mirror Books.

Thurnell-Read, T. and Parker, A. (2008a). Men, Masculinities and firefighting: Occupational identity, shop-floor Culture and Organisational Change. *Emotion, Space and Society*, 1(2), pp.127–134. doi: https://doi.org/10.1016/j.emospa.2009.03.001.

Thurnell-Read, T. and Parker, A. (2008b). Men, masculinities and firefighting: Occupational identity, shop-floor culture and organisational change. *Emotion, Space and Society*, 1(2), pp.127–134. doi: https://doi.org/10.1016/j.emospa.2009.03.001.

Titz, A., Cannon, T. and Krüger, F. (2018). Uncovering 'Community': Challenging an Elusive Concept in Development and Disaster Related Work. *Societies*, 8(3), p.71. doi: https://doi.org/10.3390/soc8030071.

Toman, R.M. (2018). *Public Sector Reforms in Greece: Uncertain Outcome of 2010 Reforms*. [online] Academia.edu. Available at: https://www.academia.edu/37598782/Public_Sector_Reforms_in_Greece_Uncertain_Outcome_of_2010_Reforms [Accessed 25 Apr. 2024].

Turkish Coalition of America (n.d.). *Greek War of Independence and Its Toll on Turks*. [online] www.tc-america.org. Available at: https://www.tc-america.org/issues-information/turkish-history/greek-war-of-independence-and-its-toll-on-turks-668.htm [Accessed 11 Apr. 2021].

Tziritis et al., (2024). *Learning from the Evros Fire: A review of the forest firefighting mechanism and recommendations for improvement*. Athens: WWF Greece.

United States Fire Administration (1992a). *Indianapolis Athletic Club Fire. USFA-TR-063*. USA: Federal Emergency Management Agency.

United States Fire Administration (1992b). *Wood Truss Roof Collapse Claims Two Firefighters, Memphis, Tennessee. USFA-TR-069*. USA: Federal Emergency Management Agency.

United States Fire Administration (1993). *Firefighter Autopsy Protocol*. USA: Federal Emergency Management Agency.

United States Fire Administration (2013). *Improving Firefighter Communications. Special Report 099*. USA: CreateSpace Independent Publishing Platform.

United States Fire Administration (2016). *Voice Radio Communications Guide for the Fire Service*. USA: Federal Emergency Management Agency.

United States Fire Administration (2019). *E/L/G 0300 Intermediate Incident Command System for Expanding Incidents, ICS 300. Student Manual*. U.S. Fire Administration.

United States Fire Administration (1999). *Developing Effective Standard Operating Procedures for Fire and EMS Departments FA-197*. USA: Federal Emergency Management Agency.

United States Fire Administration (2020). *Emergency Services Ergonomics and Wellness*. Maryland: United States Fire Administration.

van der Feyst, M. (2013). *A sure-fire Way to Stop Fireground Freelancing*. [online] FireRescue1. Available at: https://www.firerescue1.com/cod-company-officer-development/articles/a-sure-fire-way-to-stop-fireground-freelancing-QXwEm6qWf8lv6aFo/ [Accessed 25 Oct. 2022].

Van, B. (2009). *Learning from near Misses: What the National Fire Fighter Near-Miss Reporting System Means for EMS - JEMS: EMS, Emergency Medical Services - Training, Paramedic, EMT News*. [online] Jems.com. Available at: https://www.jems.com/operations/learning-near-misses-what-nati/ [Accessed 10 Oct. 2021].

Varone, C. (1996). *Fireground Radio Communications and Firefighter Safety.156*. USA: Executive Fire Officer Program.

Varone, C. (2003). *FIREFIGHTER SAFETY AND RADIO COMMUNICATION - Fire Engineering: Firefighter Training and Fire Service News, Rescue*. [online] https://www.fireengineering.com/. Available at: https://www.fireengineering.com/leadership/legal/firefighter-safety-and-radio-communication/#gref [Accessed 10 Aug. 2022].

Varoufakis, I., Patokos, T., Tserkezis, L. and Koutsopetros, C. (2011). *The Economic Crisis in Greece and Europe in 2011*. Athens: Institute of Employment. Observatory of Economic and Social Development.

VFIS (2018). *Cancer Prevention: Personal Protective Equipment (PPE) Decontamination*. [online] Available at: https://www.indfirechiefs.org/uploads/3/4/8/9/34896601/vfis-cancerprevention-ppedecontamination-0718_1__1.pdf [Accessed 25 Mar. 2023].

Vlachos, D. (2018). *Health and Safety in the Fire Service*. [online] Enotiki Agonistiki Kinisi Pyrosveston. Available at: https://www.eakp.gr/2018/04/01/%CE%B4%CE%B5%CE%BB%CF%84%CE%B9%CE%BF-%CF%84%CF%85%CF%80%CE%BF%CF%85-%CE%B3%CE%B9%CE%B1-%CF%84%CE%B7%CE%BD-%CF%80%CF%81%CE%B1%CE%B3%CE%BC%CE%B1%CF%84%CE%BF%CF%80%CE%BF%CE%B9%CE%B7%CF%83%CE%B7-%CF%84/ [Accessed 18 Jun. 2018].

Vulliamy, E. (2013). *We Cherish the Southern European lifestyle yet Scorn Southern Europeans | Ed Vulliamy*. [online] The Guardian. Available at: https://www.theguardian.com/commentisfree/2013/aug/17/northern-europe-southern-europe-attitudes.

Walker, B. (2016). *Fire Dynamics for Firefighters: Making Fire Dynamics Science Accessible to Firefighters and Practitioners at All Levels*. Hove, East Sussex: Pavilion.

Warren, T. (2014). *The Drama of Politics in the Fire Service.* [online] https://www.fireengineering.com/. Available at: https://www.fireengineering.com/leadership/the-drama-of-politics-in-the-fire-service/#gref [Accessed 15 May 2019].

Warren, T.N. (2016). *Standard Operating Procedures: The First Step to a Safer Fireground.* [online] https://www.firefighterclosecalls.com/. Available at: https://www.firefighterclosecalls.com/standard-operating-procedures-the-first-step-to-a-safer-fireground/ [Accessed 10 Oct. 2021].

Waters, M. (2014). *Responding to 'Unknown' Emergencies.* [online] *www.fireengineering.com.* Available at: https://www.fireengineering.com/technical-rescue/feu-responding-to-unknown-emergencies/#gref [Accessed 6 Nov. 2018].

Watterson, Andrew. (2015). Firefighter fatalities at fires in the UK: 2004-2013: Voices from the fireground. 10.13140/RG.2.1.3001.7529.

Welsh Government (2019). *National Framework for Fire and Rescue Services. Progress Report 2019.* Wales: Welsh Government.

Wendt, A. (1992). Anarchy is what States Make of it: The Social Construction of Power Politics. *International Organization,* [online] 46(2), pp.391–425. Available at: http://www.jstor.org/stable/2706858?origin=JSTOR-pdf.

West Midlands Fire Service (2024). *Our Community Risk Management Plan 2024 - 2027.* Birmingham: West Midlands Fire Service.

What Is the Incident Command System (ICS)? (n.d.). Available at: https://www.seapro.org/pdf_docs/ICS.Overview.pdf [Accessed 18 Jun. 2018].

Willing, L. (2012). *3 Keys to Changing Fire Service Culture.* [online] FireRescue1. Available at: https://www.firerescue1.com/fire-service-culture/articles/3-keys-to-changing-fire-service-culture-U6kiSpDo5knaUeLA/ [Accessed 11 Apr. 2021].

Winchester, N. (2022). *Review of the London Fire Brigade: Findings and Reaction*. [online] House of Lords Library. Available at: https://lordslibrary.parliament.uk/review-of-the-london-fire-brigade-findings-and-reaction/#heading-3 [Accessed 20 Apr. 2023].

Woodside, A.G. (2012). Incompetency training: Theory, Practice, and Remedies. *Journal of Business Research*, 65(3), pp.279–293. doi: https://doi.org/10.1016/j.jbusres.2011.10.025.

Workgroup Report (2024). *Firefighter Cancer Accomplishments: 4 Development of a Linear Model to Guide the Firefighter Cancer Workgroup on National Strategic Planning*. [online] Available at: https://www.usfa.fema.gov/downloads/pdf/summit/2024/2024-summit-firefighter-cancer-workgroup-report.pdf?utm [Accessed 23 Apr. 2025].

WWF Greece (2008). *A Sustainable Future from Greece's Forests: a Proposal for the Effective Improvement of Forest Protection System from Fires*. Athens: WWF Greece.

Xanthopoulos, G. (2021). S*uppression versus prevention-the disastrous forest fire season of 2021 in Greece*. [online] International Association of Wildland Fire. Available at: https://www.iawfonline.org/article/suppression-versus-prevention-the-disastrous-forest-fire-season-of-2021-in-greece/ [Accessed 5 Nov. 2022].

Yamamoto, B. (2023). *Surveying Perceptions and Understanding of Fire/EMS: What the Public Thinks and Knows*. [online] Scholar Works. Available at: https://scholarworks.calstate.edu/concern/projects/zc77sx58c [Accessed 30 Jun. 2024].

Yampolsky, M.A., Amiot, C.E. and de la Sablonnière, R. (2013). Multicultural Identity Integration and well-being: a Qualitative Exploration of Variations in Narrative Coherence and Multicultural Identification. *Frontiers in Psychology*, [online] 4(126). doi: https://doi.org/10.3389/fpsyg.2013.00126.

Yin, R.K. (2009). *Case Study research: Design and Methods*. 4th ed. Los Angeles: Sage Publications.

Yin, R.K. (2014). *Case Study research: Design and Methods*. 5th ed. Los Angeles: Sage.

Zacks, J.M. (2020). Event Perception and Memory. *Annual Review of Psychology*, 71(1), pp.165–191. doi: https://doi.org/10.1146/annurev-psych-010419-051101.

Zamora-Kapoor, A. and Coller, X. (2014). The Effects of the Crisis. *American Behavioral Scientist*, 58(12), pp.1511–1516. doi: https://doi.org/10.1177/0002764214530649.

Zimmermann, K.A. (2018). *German Culture: Facts, Customs and Traditions*. [online] Live Science. Available at: https://www.livescience.com/44007-german-culture.html [Accessed 22 Jan. 2020].

Zoakos, C. (2010). *Eye Popping Greek Corruption and the Collusion between Athens Officials and EU interests*. [online] Available at: http://www.international-economy.com/TIE_Sp10_Zoakos.pdf [Accessed 24 Aug. 2018].